Landscape Plants for the Southeast

University of South Carolina Press

LANDSCAPE PLANTS
For the Southeast

Text and Drawings by

Wade T. Batson

Published in Columbia, South Carolina by the
University of South Carolina Press

First Edition

Manufactured in the United States of America

Library of Congress Cataloging in Publication Data

Batson, Wade T.
Landscape plants for the Southeast.

Includes indexes.
1. Plants, Ornamental—Southern States. 2. Landscape
gardening—Southern States. I. Title.
SB407.B34 1984 635.9'0975 84-5267
ISBN 0-87249-433-0

CONTENTS

PREFACE

This book is to provide assistance for the identification and landscape use of native and ornamental plants generally available in the Southeast. At any given time the number of species, varieties, forms and hybrids cultivated here is unknown and probably increases year by year. An attempt to include all that are grown by homeowners, municipalities, parks, institutions, breeders, fanciers and collectors would complicate and extend this work to an unmanageable size.

Most of the plants commonly thought of as ornamentals have been collected from various parts of the earth and introduced into cultivation. Through the extensive and competitive efforts of plant breeders many species and groups of species have been developed into varieties, forms and hybrids seemingly without number—Roses, Azaleas, Honeysuckles, Syringas and Viburnums, to mention a few. Some of these subspecific types are formally recognized by Latin names; others have been given only common names. From the standpoint of the practical landscaper, differences are often negligible and the multiplicity of forms very confusing. This work recognizes only the major ones, or sometimes only major groups of these.

In this country the relatively short history of native plants in ornamental use results in only a few having been domesticated to the point of subspeciation. Quite a number, however, are commonly propagated or taken from their native habitats by dealers and others and used or offered as ornamentals.

A number of little known native species, none without landscape potential, however, are included with the hope that they may gain wider appreciation. This would result in their greater use and for those now listed as rare and endangered a better chance for survival.

Information on transplanting, fertilizing and pruning is well covered by various easily available sources and is omitted here. Insect and disease control is also omitted as both materials and methods are undergoing rapid changes.

PREFACE

[illegible]

ACKNOWLEDGMENTS

Sincere appreciation is hereby expressed to many whose interests and assistance over the years have enabled me to put together this presentation.

Special appreciation is extended to The Garden Club of South Carolina, and their many affiliates for their long standing interest in conservation and landscape beautification.

To Dr. John L. Frierson and Mr. T. Michael James of Columbia, and to Mr. Robert L. Mackintosh, Mrs. Julia Mackintosh, Mr. Robert B. McCartney and Professor Harry E. Shealy, all of Aiken, for their great interest in the use of greater variety in outdoor ornamentals and the greater use of native species.

To Mr. Robert Marvin of Summerville for his interest in preserving natural beauty in the landscape.

To Professor R. Gordon Halfacre of Clemson for his significant contributions to landscaping.

To Dr. Oscar LaBorde and Professor David H. Rembert, Jr., of Columbia for their interest in and assistance with identification.

To Professor Brooks C. Metts and Mrs. Gwyndolyn Ballentine of the Palmetto Poison Center for work with poisonous species.

To my students whose questions and observations have been an inspiration.

To many others from all walks of life for helpful inquiries and collections.

And, to my wife, Josephine, for her indulgence and support.

USES

The use of a particular kind of plant for a specific location on the landscape varies according to the physical characteristics of the spot, along with, of course, the taste of the landscaper and the kinds of plants available. A listing of some of the possible places where, or ways in which, plants may be used follows.

- accent
- arbor
- autumn color
- backdrop
- bank
- barrier
- bed
- blender
- chimney
- coastal garden
- decoration
  - flowers
  - foliage
  - fruit
- dry sites
- edging
- facing
- fence cover
- foundation
- fragrance
- framing
- free standing
- ground cover
- hanging basket
- hedge
- herb garden
- informal shrub border
- interest
- margin
- masonry
- naturalistic effect
- occasional flowering shrub
- patio
- pergola
- planter
- pot
- protection
- rock garden
- shade
- sidewalk
- single
- softening
- specimen
- street or avenue
- sun screen
- terrace
- texture of foliage
- trellis
- unit arrangement
- view screen
- wall
- waterfront
- wet spot
- wind break
- winter appeal
- woodland

SOIL

Soil provides plants with anchorage, mineral nutrients, a supply of water and aeration for their roots. Different kinds of plants require these essentials to different degrees but land plants must have some of all four. Soil is composed of mineral particles derived from the breakdown of parent rock, organic materials added or left behind by plants and animals, water and air. These essentials vary from place to place and influence or determine the species that grow there.

The size of the mineral particles forms the basis for soil texture classes; gravel, coarse, medium or fine sand, silt and clay. Texture has much to do with moisture, aeration and productivity. When one size class does not predominate in a sample but two or more are common, that sample, or soil, is referred to as a loam and may be described according to the size class that predominates as coarse sandy loam, silt loam, etc. Most soils are loams.

Soil Moisture

The ultimate source of moisture or water found in the soil is rain, sleet, snow and hail. Not all precipitation percolates into the soil. Such factors as texture of the soil, degree of slope, organic content, amount of cover or litter, rate of precipitation and amount of water in the soil to begin with influence the total amount of infiltration. Plant growth probably is more commonly affected by soil water than by any other soil factor.

A coarse textured soil having large spaces between the particles not only allows for rapid percolation but also for rapid drainage. A fine textured soil with very small spaces between the particles can easily become water-logged with little or no space for soil air.

Plant roots pick up water from the soil by means of root hairs which occur in a short zone near the tips of roots. These structures absorb directly from the spaces, when they are water-filled, and from the capillary film remaining around the particles when the spaces are empty.

When a plant is moved, usually much of the root system is lost, especially the very tender root tips and root hairs. For this reason water and mineral absorption is drastically reduced and survival of the plant threatened. What is hoped for is that new root growth, supported by reserves in the plant, will develop quickly. To better insure that this happens one can compensate for the reduced water absorption by reducing water loss. Since this loss takes place from the above ground parts, remove some of the branches or branch tips at time of transplanting, and especially with bare root specimens.

SELECTION GUIDE

Most plants are adapted to fairly specific environmental requirements (habitats), and as might be expected, to force a plant to grow outside its range of these requirements is difficult. In overcoming this, two options are open; either select a plant suited to the proposed new environment or modify the conditions of the site to suit the particular plant. The first is often easier.

Following are a few specific environmental situations along with a limited list of plants likely to succeed in each. Also, listings of hardy vines and plants developing bright autumn color are provided along with some ground cover and low shrub types.

For Seashore Planting

Common Name	*Scientific Name*
Althea	*Hibiscus*
Batis	*Batis*
Blueberry Highbush	*Vaccinium corymbosum*
Buckthorn	*Rhamnus*
Butchers Broom	*Ruscus*
Climbing Hydrangea	*Schizophragma*
Greenbrier	*Smilax*
Groundsel	*Baccharis*

Holly	*Ilex*
glabra	*glabra*
opaca	*opaca*
vomitoria	*vomitoria*
India Hawthorn	*Rhaphiolepis*
Lavender	*Lavandula*
Marsh Elder	*Iva*
Oleander	*Nerium*
Palmetto	*Sabal*
Pepper-Vine	*Ampelopsis*
Prickly-Ash	*Xanthoxylum*
Red Cedar	*Juniperus*
Rose	*Rosa*
multiflora	*multiflora*
palustris	*palustris*
Russian Olive	*Elaeagnus*
Salt-Bush	*Atriplex*
Sea Oxeye	*Borrichia*
Spanish Bayonet	*Yucca*
Spirea	*Spiraea*
Viburnum	*Viburnum*
cassinoides	*cassinoides*
dentatum	*dentatum*
tinus	*tinus*
Wax-Myrtle	*Myrica*
Wisteria	*Wisteria*

For Very Dry Sites, Often Sandy

Common Name	***Scientific Name***
Barberry	*Berberis*
Bottle-Brush	*Callistemon*
Butterfly-Bush	*Buddleja*
Dangleberry	*Gaylussacia*
Deerberry	*Polycodium*
Euonymus	*Euonymus japonica*
Gorse	*Ulex*
Groundsel	*Baccharis*
Hackberry	*Celtis*

Indigo	*Indigofera*
Jointweed	*Polygonella*
Lavender-Cotton	*Santolina*
Lead-Plant	*Amorpha*
Muscadine	*Vitis*
Myrtle	*Myrtus*
New Jersey Tea	*Ceanothus*
Oleander	*Nerium*
Pagoda Tree, Japanese	*Sophora*
Red Cedar	*Juniperus*
Russian Olive	*Elaeagnus*
St. Andrew's Cross	*Ascyrum*
Scotch-Broom	*Cytissus*
Smoke-Tree	*Cotinus*
Spanish Bayonet	*Yucca*
Sumac	*Rhus*
Tamarisk	*Tamarix*
Trailing Arbutus	*Epigaea*
Wax-Myrtle	*Myrica*
Weavers-Broom	*Spartium*
Winged Elm	*Ulmus*

For Wet Places

Common Name	***Scientific Name***
Alder	*Alnus*
Arbor-Vitae	*Thuja*
Bamboo	*Arundinaria*
Blueberry, Highbush	*Vaccinium corymbosum*
Button-Bush	*Cephalanthus*
Cornus	*Cornus*
amomum	*amomum*
stricta	*stricta*
Dog-Laurel	*Leucothoe*
Elder-Berry	*Sambucus*
Holly	*Ilex*
decidua	*decidua*
vomitoria	*vomitoria*
Loblolly-Bay	*Gordonia*

Palmetto	*Sabal*
Red-berried Greenbrier	*Smilax walteri*
Rose, Swamp	*Rosa palustris*
Rush	*Juncus*
Spice-Bush	*Lindera*
Swamp Honeysuckle	*Rhododendron*
Sweet Bay	*Magnolia*
Sweet Pepper-Bush	*Clethra*
Viburnum	*Viburnum cassinoides*
Virginia Willow	*Decodon*
Willow	*Salix*

Vines or Vine-Like Shrubs

Common Name	***Scientific Name***
Akebia	*Akebia*
Bittersweet	*Celastrus*
Boston-Ivy	*Hedera*
Carolina Jessamine	*Gelsemium*
Carolina Moonseed	*Menispermum*
Cat's Claw-Vine	*Doxantha*
Climbing Hydrangea	*Decumaria*
Climbing Hydrangea	*Schizophragma*
Confederate-Jasmine	*Trachelospermum*
Cross-Vine	*Bignonia*
Cupseed	*Calycocarpum*
English-Ivy	*Hedera*
Euonymus	*Euonymus fortunei*
Kiwi-Berry	*Actinidia*
Manettia	*Manettia*
Matrimony-Vine	*Lycium*
Moonseed	*Cocculus*
Muscadine	*Vitis*
Pepper-Vine	*Ampelopsis*
Pipe-Vine	*Aristolochia*
Poets-Jasmine	*Jasminum officinale*
Trumpet-Honeysuckle	*Lonicera sempervirens*
Trumpet-Vine	*Campsis*
Twining-Vine	*Schisandra*

Virginia Creeper	*Parthenocissus*
Virgin's-Bower	*Clematis*
Wisteria	*Wisteria*

Autumn Colors

Leaves contain four basic pigments; two greens, both chlorophylls; and two yellows, carotin and xanthophyll. Normally the greens mask the yellows. Varying amounts of each of these pigments, along with the thickness of the epidermis and cuticle, and the amount of hairs or scales on the leaf surface account for the different shades of green that are reflected by different species. This is because the chlorophyll absorbs little of the green wavelengths of the visible spectrum; therefore green is reflected. Ample nitrogen availability in the soil increases the production of chlorophyll and for that reason well fertilized plants may, because of a great abundance of chlorophyll, become a darker green than normal.

In temperate latitudes and where there is ample rainfall, nature presents an annual grandiose spectacular—autumn colors. In North America a brilliant display occurs in southeastern Canada, the northeastern United States and, because of altitude, south along the mountain chains into Georgia. This is an area of predominantly deciduous forests.

Yellow is almost always an autumn color and when shortening days and lowering temperature, particularly at night, decrease chlorophyll production well below the level required for normal replacement the greens begin to fade thereby unmasking the always-present yellows. In a few plants, such as Magnolias, the greens and yellows breakdown simultaneously and the leaves turn directly from green to dead brown. Variation, whether induced genetically or environmentally, is a characteristic of living things and accounts for not all individuals of a species changing exactly the same way or at the same time.

The reds, scarlet, and purplish colors, present only in certain species, result from the formation of a different pigment, anthocyanin. The formation of this pigment is dependent on an excessive accumulation of sugar and tannin in the leaf cells. Some Maples and Oaks are examples.

Three conditions are necessary for brilliant autumn colors. First, adequate rainfall during late summer and into the fall, so that the trees are thriving. Second, warm bright sunny days, during which sugar and other materials are manufactured in quantity. Third, cool nights with temperatures below 50°F, during which the physiological process of translocation of these materials from the leaves to stems and roots does not occur.

The period of vivid autumn colors is sometimes abbreviated by rain. The beating of the rain drops and weight of the water may cause many leaves to fall a few days early.

Some Plants for Good Autumn Colors:

Common Name	***Scientific Name***
Barberry	*Berberis*
Blackgum	*Nyssa*
Crepe-Myrtle	*Lagerstroemia*
red-flowered varieties	*red-flowered varieties*
Dogwood	*Cornus florida*
Enkianthus	*Enkianthus*
Huckleberry, Highbush	*Vaccinium corymbosum*
Maple	*Acer*
Japanese	*palmatum*
Red	*rubrum*
Sugar	*saccharum*
*Southern Sugar	*saccharum floridanum*
Nandina	*Nandina*
Oak	*Quercus*
Scarlet	*coccinea*
Turkey	*laevis*
White	*alba*
Parrotia	*Parrotia*
Persimmon	*Diospyros*
Sassafras	*Sassafras*
Sourwood	*Oxydendrum*
Spirea	*Spiraea prunifolium*
Sumac	*Rhus*
Sweetgum	*Liquidambar*
Viburnum	*Viburnum*
*acerifolium	**acerifolium*
cassinoides	*cassinoides*
prunifolium	*prunifolium*
Virginia Creeper	*Parthenocissus*
Witch-Alder	*Fothergilla*
*Very Good	**Very Good*

Ground Cover Types

Common Name	***Scientific Name***
Bugle-weed	*Ajuga*
Centella	*Centella*
Dichondra	*Dichondra*
English-Ivy	*Hedera*
Fog-Fruit	*Lippia*
Galax	*Galax*
Indian Strawberry	*Duchesnea*
Juniper	*Juniperus*
Marsh Pennywort	*Hydrocotyle*
Monkey-Grass	*Liriope spicata*
Moss-Pink	*Phlox*
Pachysandra	*Pachysandra*
Partridge-berry	*Mitchella*
Periwinkle	*Vinca*
Phlox	*Phlox*
Pink	*Dianthus*
Sandwort	*Arenaria*
Sedge	*Carex*
Spiderwort	*Tradescantia*
Stonecrop	*Sedum*
Yellow Archangel	*Lamiastrum*

Erect Types Usually Not Over 3′ High

Common Name	***Scientific Name***
Boxwood	*Buxus*
Bush-Cinquefoil	*Potentilla*
Bush-Clover	*Colutea*
Butchers-Broom	*Ruscus*
Butterfly-weed	*Asclepias*
Cast Iron-Plant	*Aspidistra*
Christmas Rose	*Helleborus*
Coral-Bean	*Erythrina*
Coral-Berry	*Symphoricarpos*
Escallonia	*Escallonia*
Lantana	*Lantana*

Lavender	*Lavandula*
Live-for-ever	*Sedum*
Loropetalum	*Loropetalum*
Marlberry	*Ardisia*
Meadow-Sweet	*Spiraea*
Minnie-Bush	*Menziesia*
Nolina	*Nolina*
Oregon-Grape	*Mahonia*
Rush	*Juncus*
St. Andrew's Cross	*Ascyrum*
Salt-Bush	*Atriplex*
Savory	*Satureja*
Sea Oxeye	*Borrichia*
Sweet-Fern	*Comptonia*
Turkeybeard	*Xerophyllum*
Spanish Bayonet	*Yucca*

HARDINESS ZONES

Every plant species is restricted to a definite temperature range. Tolerance to freezing is particularly noticeable and if this occurs injury or death may result. Such may be brought about by any or all of several internal changes. Mechanical injury may result from water crystalization causing the rupture of cell walls and tissue. Chemical injury may be sustained when freezing results in the desiccation of certain proteins and the vacuole fluid. In other instances, especially with subtropical species, death may occur before the temperature reaches freezing.

At any rate, low temperature is a very important limiting factor to range. For the area of the Southeast, the physiographic provinces may be thought of as temperature zones: (1) Coastal Plain, (2) Piedmont, and (3) Mountains. Very roughly these divisions respectively are comparable to average minimum temperature zones 9, 8 and 7 as listed elsewhere. Zones 3 and 7 experience the lowest temperatures.

Descriptions

Glossy Abelia

Abelia X grandiflora (A. chinensis X A. uniflora)

Honeysuckle Family

GROWTH HABIT: Semievergreen shrub to 6′
LEAVES: Opposite, ovate, toothed, glossy above and to 1″ or more long
FLOWERS: Pinkish, and in small clusters
FRUITS: Leathery archene
SEEDS: 1 seed
LIGHT: Lots of sun
SOIL: Any fertile type
MOISTURE: Moist
SEASONAL ASPECT: Summer flowers, bronze foliage in fall
ZONE: 1, 2, 3
USES: Hedge, shrub border, occasional shrub
CARE: Pruning to maintain size and shape
PROPAGATION BY: Cuttings
NATIVE OF: A hybrid from Chinese stock

Abeliophyllum, White Forsythia

Abeliophyllum distichum

Olive Family

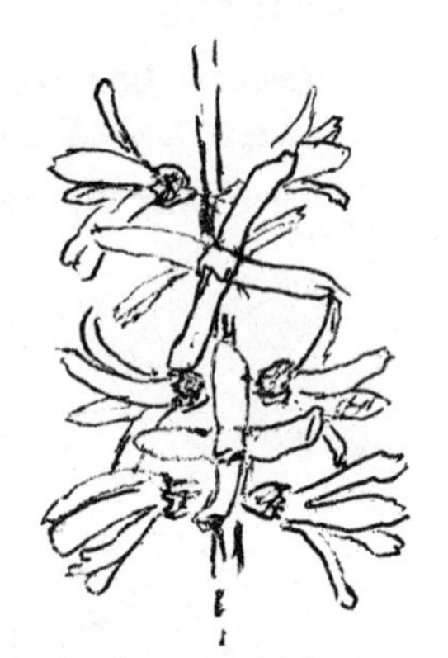

GROWTH HABIT: Much-branched, deciduous shrub to 3′ high
LEAVES: Opposite, oblong, to 2″ long and entire
FLOWERS: White, 4-parted, 1/2″ wide and very abundant
FRUITS: —
SEEDS: —
LIGHT: Full or partial sun
SOIL: Any fertile type
MOISTURE: Moist
SEASONAL ASPECT: Late winter and very early spring flowers
ZONE: 1, 2, 3
USES: A very early flowering shrub
CARE: Occasional removal of overcrowded branches
PROPAGATION BY: Summer to late fall cuttings
NATIVE OF: Korea

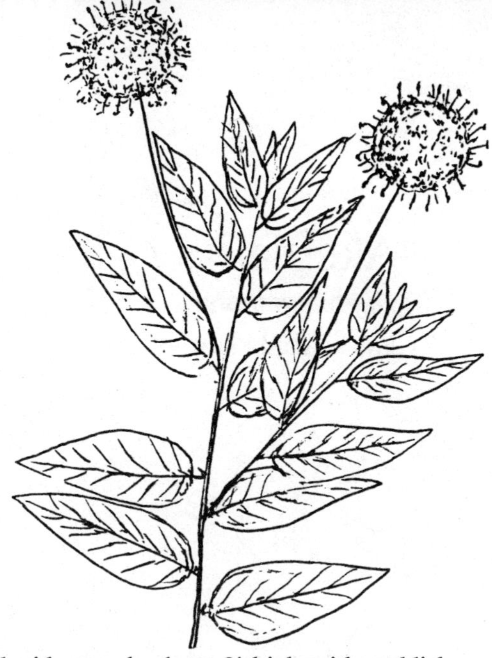

Adina

Adina rubella

Madder Family

GROWTH HABIT: Tardily deciduous shrub to 6′ high with reddish-brown pubescent twigs
LEAVES: Alternate, broadly elliptic, pointed, lustrous above and paler beneath
FLOWERS: Small and many together in long stalked heads, as in Button-Bush; lavender pink, fragrant and long lasting
FRUITS: Small capsule
SEEDS: Winged
LIGHT: Plenty of sun
SOIL: Clay loam
MOISTURE: Moist
SEASONAL ASPECT: Summer flowers
ZONE: 1, 2, 3
USES: Occasional flowering shrub, interest
CARE: None
PROPAGATION BY: Cuttings
NATIVE OF: China
OTHER: *A. pilulifera,* a larger shrub with larger leaves and greenish-yellow flowers

Akebia

Akebia quinata

Lardizabala Family

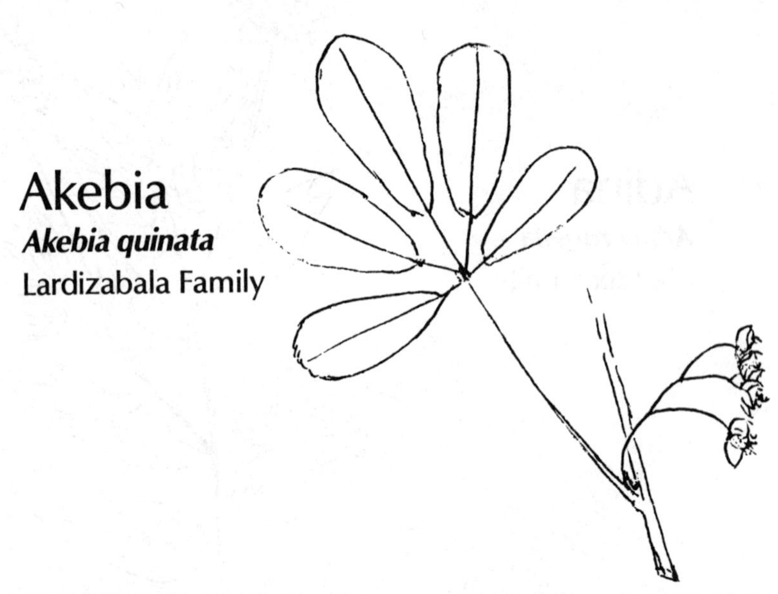

GROWTH HABIT: Tardily deciduous, woody, twining, vine climbing 10′ or more

LEAVES: Alternate, palmate with 5 leaflets, each 1-3″ long, entire and with rounded tips

FLOWERS: Brownish-purple, fragrant, male and female separate but on same raceme

FRUITS: 2-8 carpels becoming fleshy, purplish pods at maturity

SEEDS: Few per pod, usually not formed without hand pollination

LIGHT: Partial shade

SOIL: Any good loam

MOISTURE: Moist to dry

SEASONAL ASPECT: None

ZONE: 1, 2, (3)

USES: Fence, trellis, tree or bushes, arbor or pot

CARE: None

PROPAGATION BY: Seedlings or sprouts

NATIVE OF: Japan and China

Black Alder

Alnus serrulata

Birch Family

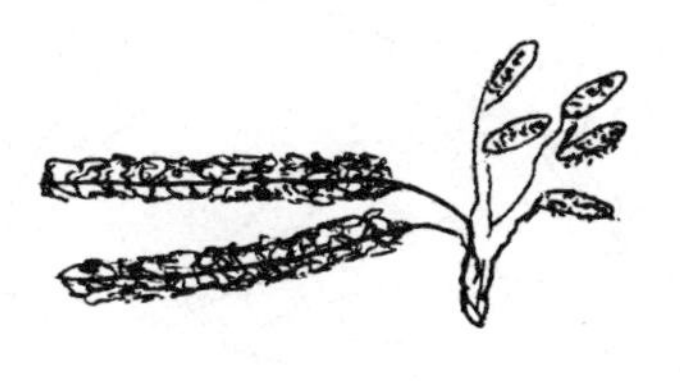

GROWTH HABIT: Deciduous shrub to 10′ high
LEAVES: Alternate, veiny toothed, dark green and somewhat pubescent
FLOWERS: Male in drooping yellow catkins in late winter; female as tiny red cones
FRUITS: Woody cone-like structure 1″ long
SEEDS: With corky thickening
LIGHT: Partial or full sun
SOIL: Silt, clay and sand
MOISTURE: Wet or moist
SEASONAL ASPECT: Yellow catkins in late winter
ZONE: 1, 2, 3
USES: Wet margins
CARE: None
PROPAGATION BY: Seedlings
NATIVE OF: e. United States

Alexandrian-Laurel

Danae racemosa

Lily Family

GROWTH HABIT: Evergreen shrub related to Asparagus and Butchers Broom, 2-4′ high with slender smooth stems
LEAVES: Minute scales at base of leaf-like stem sections (cladophylls). These sections are alternate and look and function as leaves.
FLOWERS: Tiny, yellowish and 4-6 in terminal racemes
FRUITS: Red berry 1/2″ wide
SEEDS: 2
LIGHT: Partial shade
SOIL: Fertile sandy loam
MOISTURE: Moist
SEASONAL ASPECT: None
ZONE: 1, 2, 3
USES: Interest or unit arrangement
CARE: None
PROPAGATION BY: Seeds
NATIVE OF: Asia Minor

Althea, Rose-of-Sharon

Hibiscus syriacus

Mallow Family

GROWTH HABIT: Large, hardy deciduous shrub to 15′
LEAVES: Ovate in outline but mostly 3-lobed, coarsely serrate, glabrous and 2-3″ long
FLOWERS: Showy, open bell shaped, 2-3″ wide, rose, purple, white or bluish
FRUITS: 1″ long 5-celled capsule
SEEDS: Numerous
LIGHT: Partial sun
SOIL: Tolerant
MOISTURE: Moist
SEASONAL ASPECT: Summer and autumn flowers
ZONE: 1, 2, 3
USES: Flowering shrub, background shrub
CARE: None
PROPAGATION BY: Seedlings
NATIVE OF: e. Asia
OTHER: Several varieties cultivated including doubled forms.
Chinese Hibiscus, *H. rosa-sinensis*; similar but to 30′ high.

American Hornbeam

Ostrya virginiana

Beech Family

GROWTH HABIT: Deciduous tree with shredded bark
LEAVES: Alternate, ovate, doubly-toothed and to 4″ long
FLOWERS: Male and female in separate catkins, no petals
FRUITS: In 2″ long clusters, nutlets in enlarged bladder-like envelopes, forming short spikes
SEEDS: 1
LIGHT: Partial shade
SOIL: Rich loam
MOISTURE: Moist
SEASONAL ASPECT: None
ZONE: (1), 2, 3
USES: Lawn, park, street, golf course, specimen
CARE: None
PROPAGATION BY: Seedlings
NATIVE OF: e. United States

Anise

Illicium anisatum

Magnolia Family

GROWTH HABIT: Evergreen shrub or small tree
LEAVES: Alternate, aromatic, thick, smooth, 3-6″ long
FLOWERS: Greenish, yellowish or purplish, 1-2″ wide and axillary
FRUITS: Several in a ring around center of flower
SEEDS: 1 per fruit
LIGHT: Shade or partial sun
SOIL: Fertile loam
MOISTURE: Moist
SEASONAL ASPECT: None
ZONE: 1, 2
USES: Occasional planting, shrub border or interest plant
CARE: None
PROPAGATION BY: Cuttings
NATIVE OF: s.e. United States
OTHER: *I. floridanum* has reddish flowers.

Five-leaved Aralia

Acanthopanax sieboldianus

Ginseng Family

GROWTH HABIT: Tardily deciduous shrub to 7′ high with arching, usually prickly branches
LEAVES: Alternate, digitate with 5-7 leaflets each to 2″ long
FLOWERS: Male and female on different plants, both small and in pendant clusters about 1″ wide, greenish-white
FRUITS: Small black berry
SEEDS: Few
LIGHT: Sun or shade
SOIL: Any good type
MOISTURE: Moist to dry
SEASONAL ASPECT: None
ZONE: 1, 2, 3
USES: Background, border, specimen
CARE: Occasional clipping
PROPAGATION BY: Mostly by cuttings
NATIVE OF: Japan
OTHER: A hardy shrub cultivated for its foliage. Will grow in sun or shade, as well as in the city. Only staminate plants are usually offered. A form with white-edged leaves is also cultivated.

Arbor-Vitae

Thuja spp.

Cypress Family

GROWTH HABIT: Aromatic evergreen tree

LEAVES: Opposite, scale-like or linear in some varieties, color varies greatly, or may be variegated

FLOWERS: Minute; male yellow; female as 6-8 opposite scales in pairs

FRUITS: Woody cones 1/2 - 1″ long

SEEDS: 2 per scale; winged or wingless

LIGHT: Full or partial sun

SOIL: Fertile loam

MOISTURE: Moist

SEASONAL ASPECT: None

ZONE: 1, 2, 3

USES: Various, depending on whether specimen is tree or shrub

CARE: Little, if any

PROPAGATION BY: Seedlings and cuttings

NATIVE OF: See below

OTHER: *T. occidentalis,* American Arbor-Vitae, n.e. United States south to N.C.

T. orientalis, Oriental Arbor-Vitae, n. China, Korea, branchlets vertical; seeds wingless.

Many cultivated forms of both species.

American or White Ash

Fraxinus americana

Olive Family

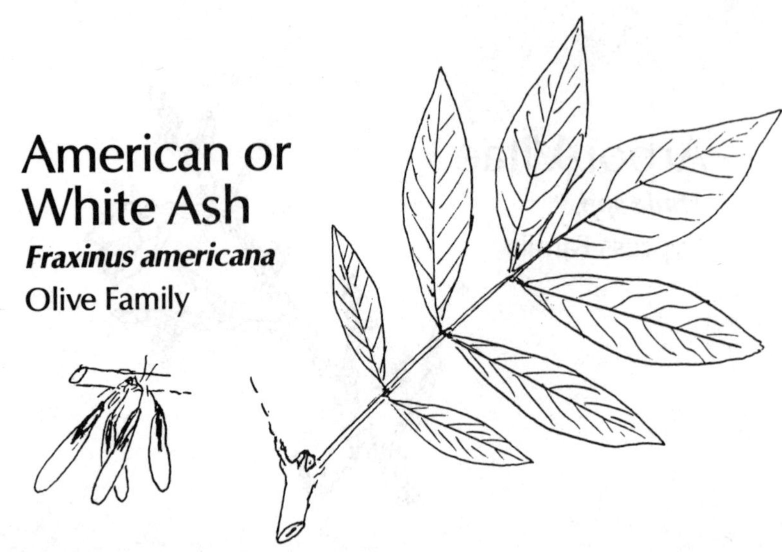

GROWTH HABIT: Large, deciduous tree with finely checkered corky bark and hard wood, long important for handles and baseball bats
LEAVES: Opposite, pinnate with usually 7-leaflets which are whitish or pale beneath
FLOWERS: Small greenish, without petals and before the leaves
FRUITS: A slender, long-winged nutlet (samara) 1″ or more long
SEEDS: 1
LIGHT: Sun or some shade
SOIL: Fertile, silty or sandy loam or alluvium
MOISTURE: Moist
SEASONAL ASPECT: None
ZONE: 1, 2, 3
USES: Freestanding, shade
CARE: None
PROPAGATION BY: Seedlings
NATIVE OF: e. United States
OTHER: Red Ash, *F. pennsylvanica*; e. United States; similar but with leaflets greenish-beneath. Also an important timber tree.
European Ash, *F. excelsior*; cultivated in many forms—dwarf, weeping, variegated, etc.

Aucuba

Aucuba japonica

Dogwood Family

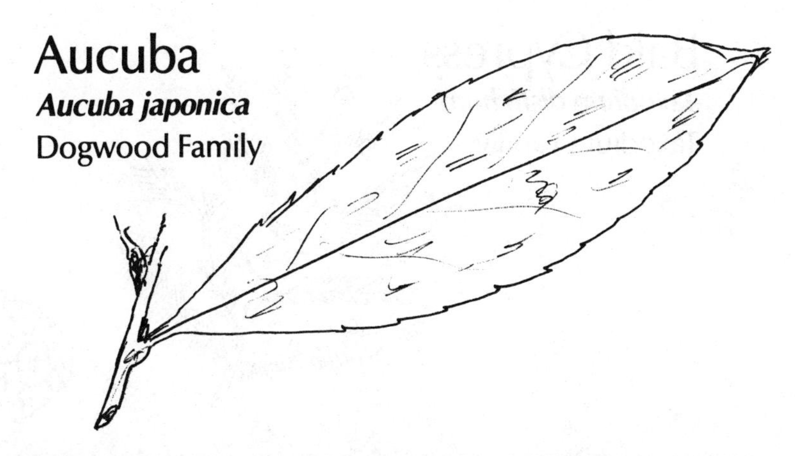

GROWTH HABIT: Stout evergreen shrub to 8′ high
LEAVES: Elliptic to oblanceolate, shallowly toothed, to 6″ long, dark green and shining or mottled with yellow
FLOWERS: Male and female on different plants but similar, small purplish and in terminal clusters
FRUITS: Red, 1/2″ long
SEEDS: 1
LIGHT: Shade
SOIL: Most any type if fertile
MOISTURE: Moist
SEASONAL ASPECT: None
ZONE: 1, 2, 3
USES: Foundation, interest, backdrop for low plants
CARE: Occasional pruning
PROPAGATION BY: Cuttings or seedlings
NATIVE OF: Japan, China and Himalayas

Bald Cypress

Taxodium distichum

Taxodium Family

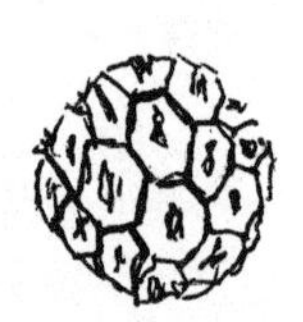

GROWTH HABIT: Deciduous conifer with single trunk and in wet sites develops a fluted base and "knees"
LEAVES: 2-ranked, flat-linear or needle-like and to 1/2" long; leaves and branchlets deciduous
FLOWERS: Males are catkin-like in terminal drooping clusters; female of imbricated scales
FRUITS: Woody cone about 1" long with thick-ended scales
SEEDS: Winged
LIGHT: Partial or full sun
SOIL: Silt or clay
MOISTURE: Wet or moist
SEASONAL ASPECT: Summer foliage
ZONE: 1, 2
USES: Freestanding, interest, wet borders
CARE: None
PROPAGATION BY: Seedlings
NATIVE OF: s.e. United States

Bamboo, Switch-Cane, Native

Arundinaria gigantea

Grass Family

GROWTH HABIT: Semievergreen perennial, our native woody grass, 2-6 or more feet high, rarely to 20

LEAVES: Linear, pointed, 1/4 - 1″ wide and 4-8″ long and with petiole that forms a sheath around stem

FLOWERS: Green spikelets disposed mostly in axillary or terminal panicles

FRUITS: Formed infrequently

SEEDS: 1 per spikelet

LIGHT: Partial sun or shade

SOIL: Silty or sandy loam

MOISTURE: Moist

SEASONAL ASPECT: None

ZONE: 1, 2, 3

USES: Along wet margins

CARE: Occasional containment because of spread by underground rhizomes

PROPAGATION BY: Underground rhizomes

NATIVE OF: e. United States

OTHER: Species includes what was previously separated as *A. tecta.*

Bamboo, Oriental

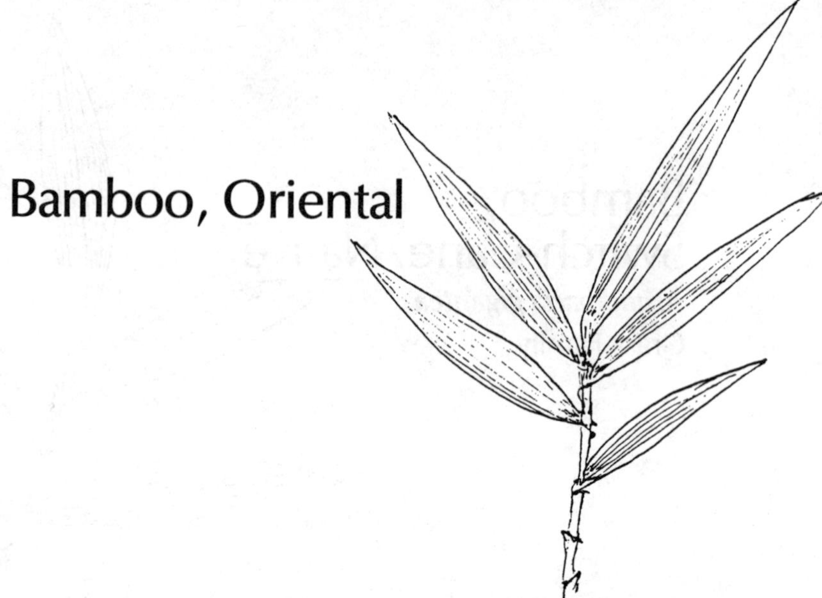

There are more than 700 of these woody grasses in the world, some attaining more than 100 feet in height. The following small ones may be available as ornamentals.

Arundinaria variegata, Dwarf Whitestripe Bamboo; to 3′ high and a rapid spreader (in good soil) by underground rhizomes. Zones 1, 2, 3.

Arundinaria viridi-striata; leaves with yellow stripes in spring, becoming green in summer, shade-loving to 2′ high. Zones 1, 2.

Phyllostachys aureosulcata; to 30′ tall, young shoots edible. Zones 1, 2.

Phyllostachys nuda; to 30′ tall; a running and spreading bamboo with very edible young shoots. Zones 1, 2.

Sasa disticha; Dwarf Fernleaf Bamboo; to 3′ high; a good ground cover. Zones 1, 2.

Sasa palmata; to 6′ high; leaves to 1′ long; a vigorous spreader. Zones 1, 2, 3.

Banana-Shrub

Michelia figo (fuscata)
Magnolia Family

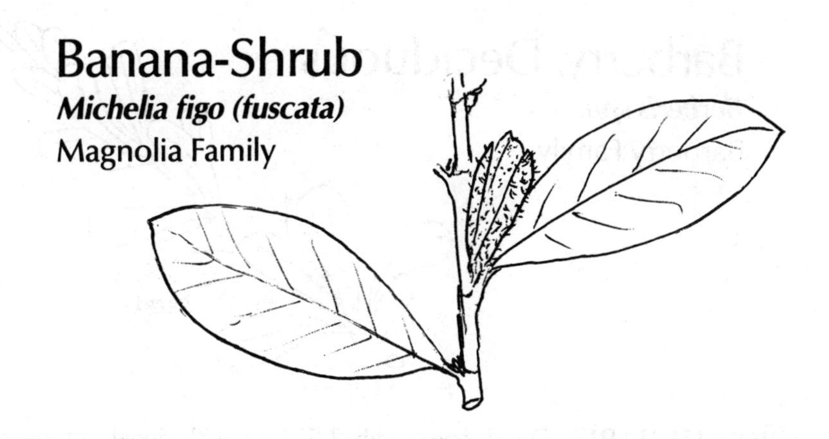

GROWTH HABIT: Large, much branched evergreen shrub with brown-hairy twigs
LEAVES: Elliptic, 2-4″ long, short pointed and dark glossy above
FLOWERS: Brownish yellow, 1″ or more long and with the fragrance of bananas
FRUITS: Several small folicles per flower
SEEDS: 1 per folicle
LIGHT: Full or lots of sun
SOIL: Fertile
MOISTURE: Moist
SEASONAL ASPECT: Fragrance of the flowers in late spring
ZONE: 1, 2
USES: Occasional shrub, informal shrub border
CARE: None
PROPAGATION BY: Seedlings and cuttings
NATIVE OF: China

Barberry, Deciduous

Berberis spp.

Barberry Family

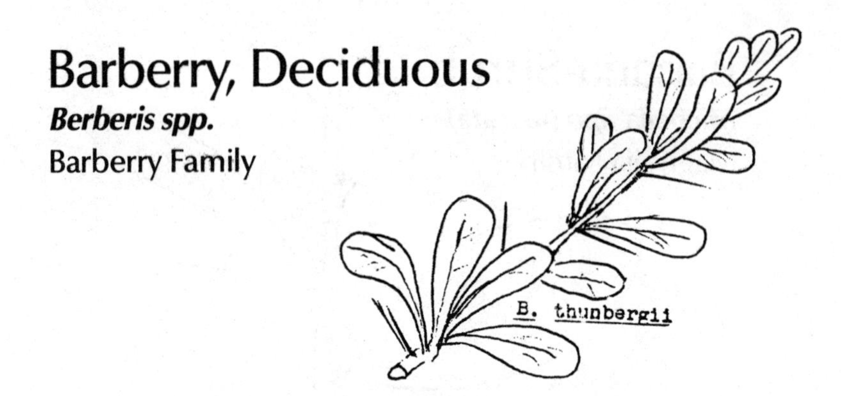

GROWTH HABIT: Deciduous shrub 2-5′ high with deeply grooved glabrous spiny branches

LEAVES: 1/2-1 1/4″ long, obovate to spoon-shaped, entire and crowded at nodes toward branch tips

FLOWERS: 1-3 per leaf axil along whole length of stem and pale yellow

FRUITS: Bright red long persisting berries

SEEDS: 1-few per berry

LIGHT: Partial or full sun

SOIL: Fertile sandy loam

MOISTURE: Moist

SEASONAL ASPECT: Bright red berries in fall and winter

ZONE: 1, 2, 3

USES: Hedge, border, unit arrangement or winter appeal

CARE: May be clipped to maintain desired size and shape

PROPAGATION BY: Cuttings

NATIVE OF: See below

OTHER: Wild Barberry, *B. canadensis*; e. N.A.; leaves toothed; spines 3-branched.

B. gilgiana; China; to 5′ high, spines unbranched, yellow flowers and long persisting blood-red berries.

Common Barberry, *B. vulgaris*; Europe, E. Asia, and naturalized in e. United States; host of Wheat Rust, now practically and purposely exterminated. The USDA has prohibited use of many species because of their lack of resistance to Wheat Rust.

Korean Barberry, *B. koreana*; a larger bush otherwise similar to the one below.

Japanese Barberry, *B. thunbergii*; dense, low shrub with leaves broadest toward the tips, simple spines and long persisting bright red berries.

Barberry, Evergreen

Berberis spp.

Barberry Family

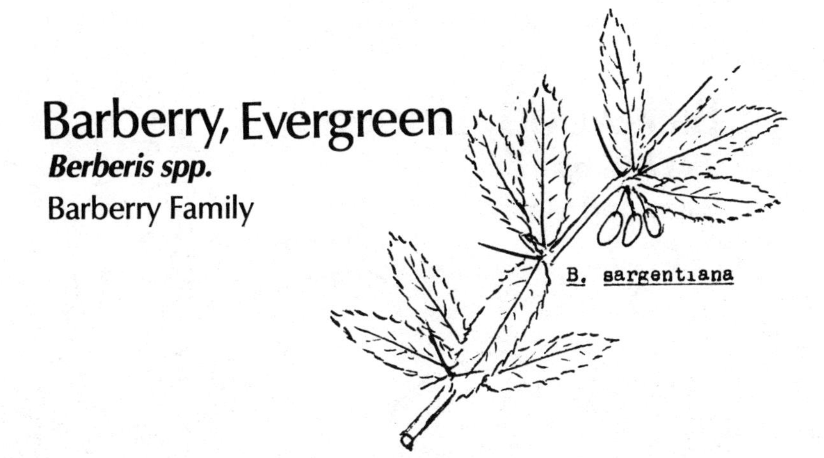

GROWTH HABIT: Evergreen shrubs to 6′ tall, branches armed with 3-branched spines
LEAVES: To 3″ in length, elliptic-oblong, leathery in texture and with spine-tipped teeth
FLOWERS: Yellow and in axillary clusters of 3-7
FRUITS: Black or bluish berry about 1/2″ long
SEEDS: Usually 1
LIGHT: Full sun or some shade
SOIL: Sandy loam
MOISTURE: Moist
SEASONAL ASPECT: None
ZONE: 1, 2, 3
USES: Hedge, unit arrangement or occasional shrub
CARE: Some pruning to maintain size and shape
PROPAGATION BY: Cuttings
NATIVE OF: China
OTHER: *B. X stenophylla*; hybrid between *B. darwinii* and *B. empetrifolia*; black berries, thorny twigs and narrow leaves to 1″ long; good hedge plant.

Winter Barberry; *B. julianae*; similar to *B. sargentiana*; but with more flowers per cluster.

B. sargentiana; China; to 6′ high, twigs with 3-branched spines; leaves to 4″ long, dark green and very spiny toothed; flowers yellow, fruit black.

B. triacanthophylla; China; to 4′ high; twigs with 3-branched spines; leaves to 2″ long and spiny toothed; flowers pinkish-white; berries black.

Basswood, Linden

Tilia heterophylla

Basswood Family

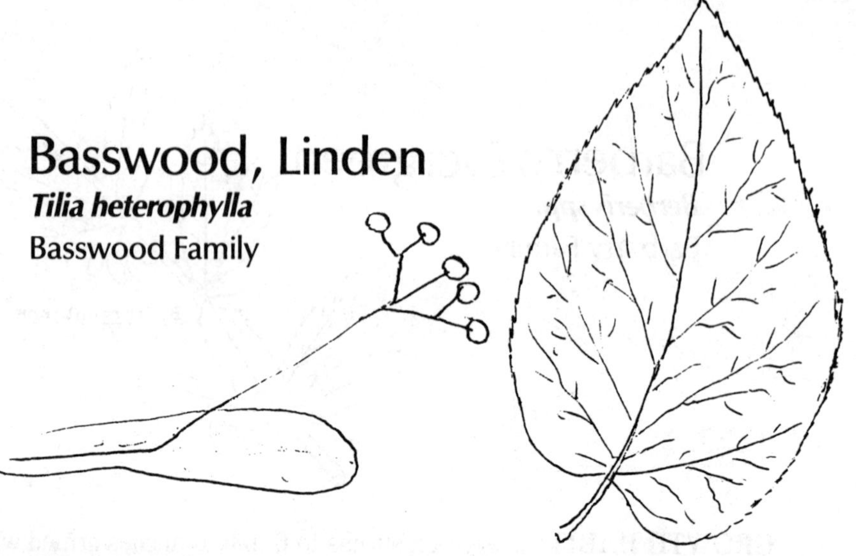

GROWTH HABIT: Large deciduous tree with glabrous twigs
LEAVES: Cordate, may be unequal sided at base, to 5″ or more long, toothed, and white or brownish beneath with densely appressed hairs
FLOWERS: 1/2″ wide, perfect, 5-parted and 3-15 together, raised on bracted peduncle
FRUITS: Indehiscent, globose, nut-like and 1/4″ wide
SEEDS: 1-3
LIGHT: Full sun
SOIL: Fertile loam
MOISTURE: Moist
SEASONAL ASPECT: None
ZONE: 1, 2
USES: Freestanding, specimen, background, shade
CARE: None
PROPAGATION BY: Seedlings
NATIVE OF: e. United States
OTHER: *T. caroliniana*; smaller tree, smaller leaves and hairy twigs.

Batis

Batis maritima

Batis Family

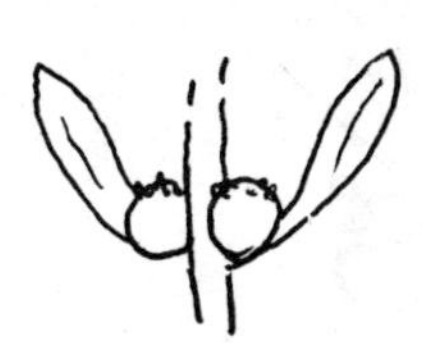

GROWTH HABIT: Succulent, semievergreen shrub with trailing stems rooting at the nodes forming colonies
LEAVES: Opposite, linear, fleshy, sessile and about 1″ long
FLOWERS: Male and female on separate plants, both small and in short cone-like groups
FRUITS: 2-8 together
SEEDS: 1
LIGHT: Full sun
SOIL: Brackish marsh, often with *Salicornia*
MOISTURE: Intertidal
SEASONAL ASPECT: None
ZONE: 1
USES: Marsh land border, salt flat
CARE: None
PROPAGATION BY: Rooted stems
NATIVE OF: s.e. United States coast

Bay, Sweet Bay

Laurus nobilis

Laurel Family

GROWTH HABIT: Small evergreen tree, usually kept as shrub
LEAVES: Stiff, dark green, fragrant when crushed, entire and to 3″ long
FLOWERS: Axillary, small, yellowish and inconspicuous, in early spring
FRUITS: Dark purple berry
SEEDS: Few
LIGHT: Partial sun
SOIL: Rich loam
MOISTURE: Moist
SEASONAL ASPECT: None
ZONE: 1, 2
USES: Occasional shrub, pot
CARE: Usually sheared to desired size and shape
PROPAGATION BY: Cuttings
NATIVE OF: Mediterranean region
OTHER: This is the Laurel of history, the Bay of Commerce and a garden plant since ancient times. Several varieties exist.

Loblolly Bay

Gordonia lasianthus

Tea Family

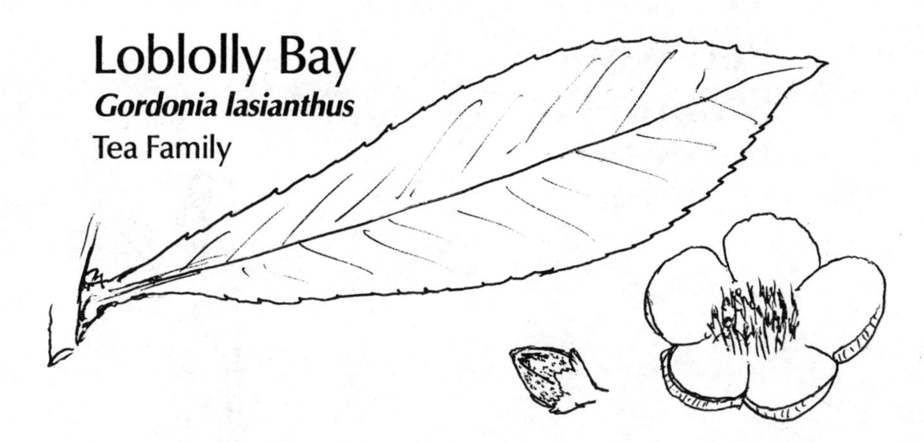

GROWTH HABIT: Evergreen tree with light gray checkered bark
LEAVES: Alternate, oblanceolate, mostly shallowly toothed, 4-7″ long, smooth
FLOWERS: White, on short pedicel from leaf axil, slightly fragrant and 3″ wide
FRUITS: Woody capsules about 1/2″ long
SEEDS: Small and winged
LIGHT: Partial shade
SOIL: Clay or silt loam with much humus
MOISTURE: Moist to wet
SEASONAL ASPECT: Flowers in summer
ZONE: 1, 2
USES: Freestanding as tree or may be pruned as shrub
CARE: Adequate moisture
PROPAGATION BY: Seedlings and cuttings
NATIVE OF: s.e. United States

Red-Bay, Swamp-Bay

Persea borbonia

Laurel Family

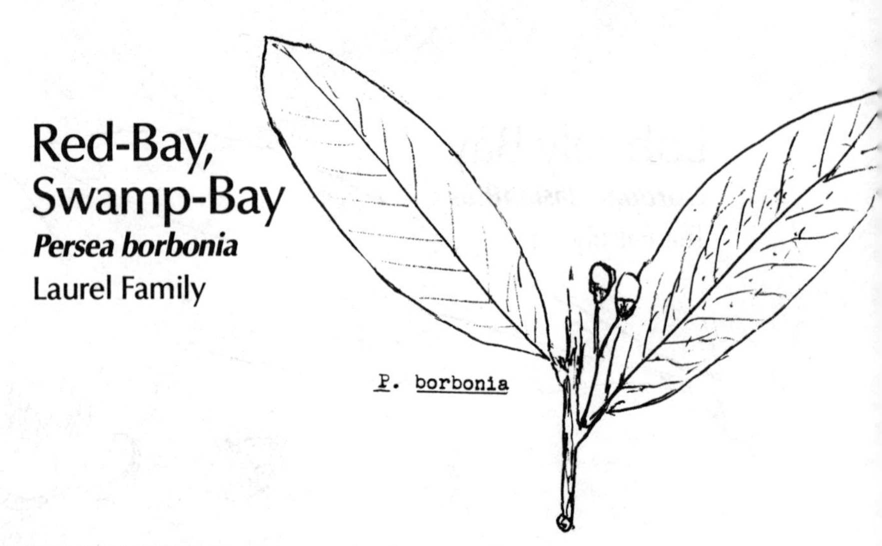

GROWTH HABIT: Evergreen shrub or small tree
LEAVES: Thick, to 6″ long, aromatic when crushed, elliptic and mostly pubescent beneath
FLOWERS: Pale yellow, small and in axillary clusters
FRUITS: Blue or black, almost 1/2″ long, calyx fleshy and persistent
SEEDS: 1
LIGHT: Half shade
SOIL: Silty or clay loam
MOISTURE: Moist to wet
SEASONAL ASPECT: None
ZONE: 1, 2, 3
USES: Use as shrub or small tre
CARE: None
PROPAGATION BY: Seedlings
NATIVE OF: s.e. United States
OTHER: The grocery store Avocado is *Persea americana* and is cultivated in the warmest parts of the United States, and sometimes as a potted curiosity elsewhere.

Beauty-Berry

Callicarpa americana

Vervain Family

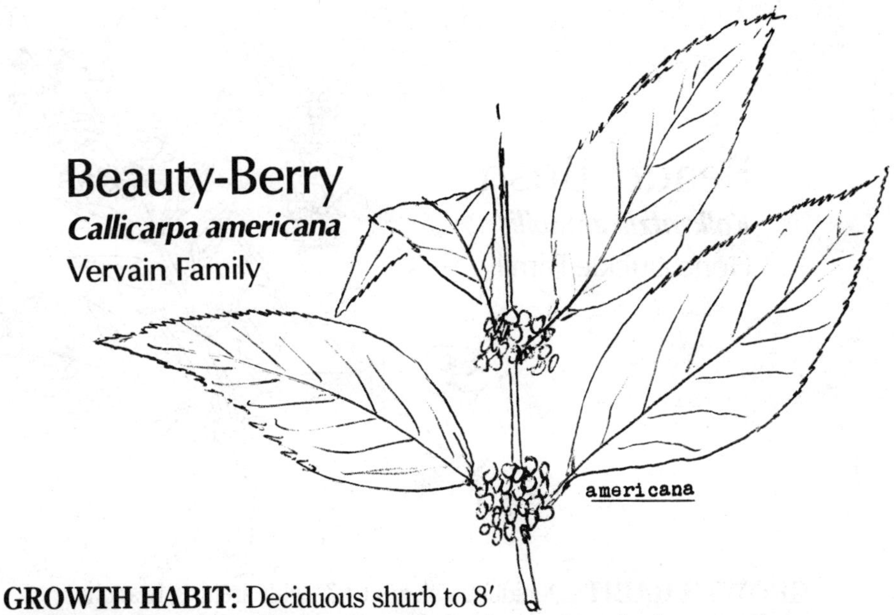

GROWTH HABIT: Deciduous shurb to 8′
LEAVES: Opposite, elliptic, somewhat veiny and toothed
FLOWERS: Small, pale lilac and in dense axillary clusters
FRUITS: Small, purplish and in dense axillary clusters
SEEDS: 1 per fruit
LIGHT: Shifting shade
SOIL: Porus soil
MOISTURE: Moist
SEASONAL ASPECT: Fruits in late summer
ZONE: 1, 2, 3
USES: Woodland plantings or as an occasional shrub in a shrub border
CARE: Prune back to near ground level each winter
PROPAGATION BY: Seeds and cuttings
NATIVE OF: s.e. United States
OTHER: Types with white fruit are sometimes available. *C. dichotoma (C. purpurea)*; China; twigs rough pubescent. *C. japonica*; Twigs smooth (both introduced species have flowers and fruits with longer stalks).

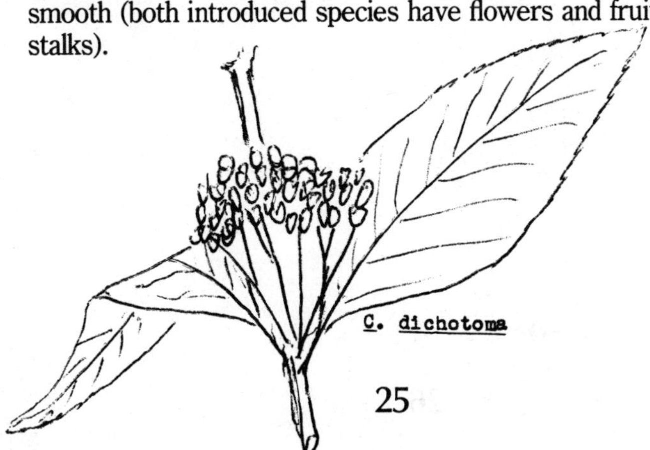

Beauty-Bush

Kolkwitzia amabilis

Honeysuckle Family

GROWTH HABIT: Deciduous shrub to 7′ with arching branches
LEAVES: Opposite, ovate, slightly toothed and to 3″ long
FLOWERS: 1/2″ long, pink with yellow center and in clusters, sepals and flower stalks bristly
FRUITS: 1/4″ long, bristly achene
SEEDS: 1
LIGHT: Partial shade
SOIL: Fertile
MOISTURE: Moist
SEASONAL ASPECT: Early summer flowers, leaves reddish in autumn
ZONE: 1
USES: Shrub border, hedge, specimen
CARE: Protect from cold
PROPAGATION BY: Seeds or cuttings
NATIVE OF: China

Beech, Beechnut

Fagus grandifolia

Beech Family

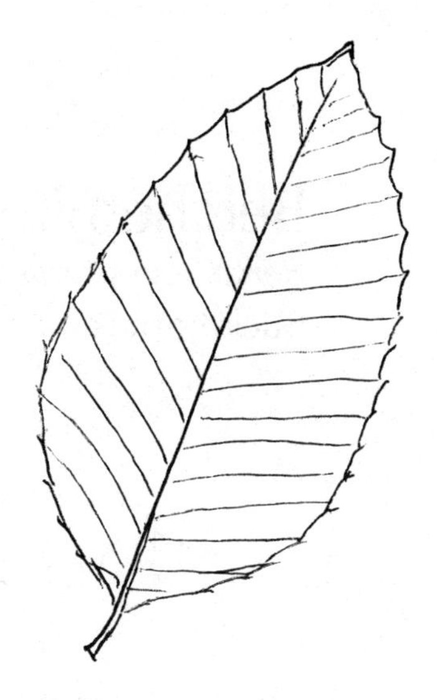

GROWTH HABIT: Large deciduous tree with smooth gray "elephant skin" bark

LEAVES: 2-4" long, dentate, straight veined

FLOWERS: Male in globose pendulous catkins, female small and inconspicuosu

FRUITS: Husk with soft coarse spines enclosing two nuts

SEEDS: 3-sided and a great favorite of birds and squirrels

LIGHT: Partial shade

SOIL: Fertile loam

MOISTURE: Moist

SEASONAL ASPECT: Winter when the papery-brown leaves are still on tree and the elephant-skin bark shows up better

ZONE: 1, 2, 3

USES: Freestanding, background

CARE: None

PROPAGATION BY: Seedlings

NATIVE OF: e. United States

OTHER: European Beech, *F. sylvestris*; available in weeping and purple-leaved forms.

Berchemia

Berchemia scandens

Buckthorn Family

GROWTH HABIT: Deciduous twining woody vine to 4″ in diameter; bark smooth
LEAVES: Alternate, elliptic, bluntish and to 2″ long
FLOWERS: Minute, greenish and terminal or nearly so
FRUITS: Black drupes 1/4″ long
SEEDS: 1
LIGHT: Partial shade
SOIL: Silty loam
MOISTURE: Moist to wet
SEASONAL ASPECT: None
ZONE: 1, 2
USES: Specimen
CARE: None
PROPAGATION BY: Seeds, seedlings
NATIVE OF: s.e. United States

River Birch, Seven-bark

Betula nigra

Birch Family

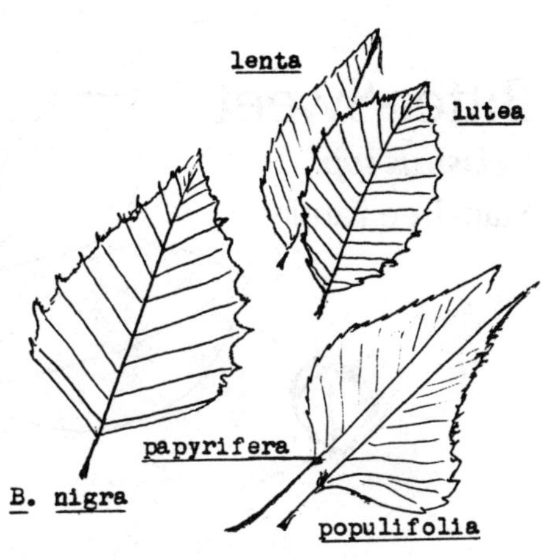

GROWTH HABIT: Good sized deciduous tree with reddish-brown bark peeling off continuously in paper-thin layers
LEAVES: 1-3″ long, somewht angular and doubly toothed
FLOWERS: Male in catkins, female very small and terminal
FRUITS: Woody, cone-like and about 1″ long
SEEDS: Winged
LIGHT: Partial or full sun
SOIL: Silty or sandy loam
MOISTURE: Moist or wet
SEASONAL ASPECT: Bark is attractive in winter
ZONE: 1, 2, 3
USES: Freestanding, interest, quick shade, low, wet margin or average situation
CARE: None
PROPAGATION BY: Seedlings
NATIVE OF: e. United States
OTHER: Yellow Birch, *B. lutea*; mountains; bud scales hairy; bark brown; leaves and twigs fragrant.
Black, Sweet or Cherry Birch, *B. lenta*; mountains; bud scales smooth; bark brown; leaves and twigs with wintergreen fragrance.
Paper Birch, *B. papyrifera*; n.e. United States; leaves with some pubescence beneath; bark white or light gray.
White Birch, *B. populifolia*; n.e. United States; small tree; leaves smooth beneath; bark gray.

Bitter-Sweet

Celastrus spp.

Staff-Tree Family

GROWTH HABIT: Deciduous, twining shrubs, to 20′ or more
LEAVES: Alternate, broadly ovate to narrowly oblong, pointed and shallowly toothed
FLOWERS: Small, greenish and 5-parted
FRUITS: 3-lobed, 3-valved capsule
SEEDS: 1-2 seeds per cell enclosed in red outer covering
LIGHT: Full or partial sun
SOIL: Tolerant of most fertile types
MOISTURE: Moist
SEASONAL ASPECT: Bright colored fruit in winter
ZONE: 1, 2, 3
USES: Wall, trellis, fence, tree
CARE: Some pruning may be desired
PROPAGATION BY: Seeds and cuttings
NATIVE OF: See below
OTHER: Oriental Bittersweet, *C. orbiculatus*; China and Japan; flower in axillary arrangements.
American Bittersweet, *C. scandens*; Quebec to North Carolina; flowers in terminal arrangements.

Black-Gum

Nyssa sylvatica

Nyssa Family

GROWTH HABIT: Large deciduous tree with extremely hard-to-split wood

LEAVES: 3-4″ long, obovate to oval, entire or remotely toothed, smooth and shining above

FLOWERS: Unisexual, male and female on same tree; small, greenish and in stalked clusters

FRUITS: Blue drupes almost 1/2″ long

SEEDS: 2

LIGHT: Full or partial sun

SOIL: Tolerant

MOISTURE: Moist

SEASONAL ASPECT: Purplish red autumn foliage and long lasting blue berries

ZONE: 1, 2, 3

USES: Freestanding, background

CARE: None

PROPAGATION BY: Seedlings

NATIVE OF: e. United States

OTHER: Formerly much used for wheels and rollers. *N. biflora,* s.e. United States; smaller and restricted to wet sites with frequent shallow water.

Tupelo; *N. aquatica*; s.e. United States; larger tree, with larger leaves and fruits. A part of Cypress-Gum Swamp and where water stands deep much of the time.

Black Ti-Ti

Cliftonia monophylla

Cyrilla Family

GROWTH HABIT: Evergreen shrub or small tree
LEAVES: Alternate, elliptic, entire, pale beneath and to 2 1/2″ long
FLOWERS: Small, whitish, 5 parted, fragrant
FRUITS: Winged nutlets
SEEDS: 1
LIGHT: Partial shade
SOIL: Heavy, acid types
MOISTURE: Wet
SEASONAL ASPECT: None
ZONE: 1
USES: Specimen, occasional shrub in wet places
CARE: None
PROPAGATION BY: Seeds and seedlings
NATIVE OF: Lower coastal plain, S.C. - Miss.
OTHER: An important honey producer

Bladdernut

Staphylea trifoliate

Bladdernut Family

GROWTH HABIT: Slender, deciduous shrub to 8′
LEAVES: Opposite, trifoliate, leaflets oval and finely toothed
FLOWERS: White, 1/3″ long and in drooping terminal panicles
FRUITS: Papery walled inflated 3-celled capsule to 1 1/2″ long
SEEDS: 1 or 2 per cell, light brown and shiny
LIGHT: Shade loving, plant under high canopy
SOIL: Fertile, alluvial soil, preferably near stream or lake margin
MOISTURE: Moist
SEASONAL ASPECT: None
ZONE: 1, 2, 3
USES: Specimen shrub
CARE: None
PROPAGATION BY: Seeds and seedlings
NATIVE OF: e. United States
OTHER: One or more similar but introduced species are sometimes available.

Bluebeard, Blue Spires

Caryopteris X clandonensis
(C. incana X C. mongholica)

Vervain Family

GROWTH HABIT: Late flowering deciduous shrub to 5′ high
LEAVES: Broadly lanceolate, somewhat toothed, to 2 1/2″ long
FLOWERS: Blue, 1/2″ long
FRUITS: Dry, capsule
SEEDS: 4
LIGHT: Sun or partial shade
SOIL: Sandy loam
MOISTURE: Moist
SEASONAL ASPECT: Late summer flowers
ZONE: 1, 2, 3
USES: As an accent plant for some of the late yellow flowering species
CARE: Prune heavily in early spring
PROPAGATION BY: Seeds
NATIVE OF: China, Mongolia

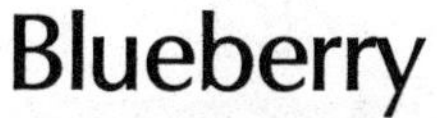

Blueberry

Vaccinium*—deciduous species, except *V. crassifolium

Heath Family

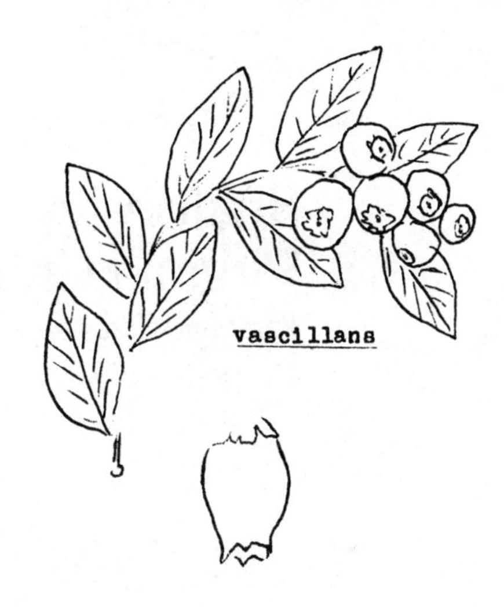

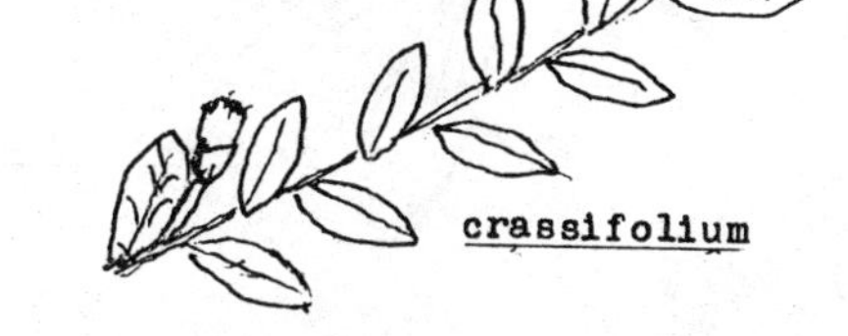

GROWTH HABIT: Mostly low to medium, deciduous shrubs
LEAVES: Alternate, entire, less than 3″ long and sometimes finely toothed
FLOWERS: White or pinkish, 1/4″ long or more
FRUITS: Powdery-blue or black berries, juicy and delicious
SEEDS: Several
LIGHT: Partial shade
SOIL: Quite tolerant
MOISTURE: Moist
SEASONAL ASPECT: Spring flowers, summer fruits
ZONE: 1, 2, 3 — see below
USES: Occasional shrub, interest
CARE: None
PROPAGATION BY: Seedlings, colony division
NATIVE OF: e. United States
OTHER: Lowbush Huckleberry, *V. tenellum*; zone 1, 2.
Lowbush Blueberry, *V. vacillans*; zone 2, 3.
Highbush Huckleberry, *V. atrococcum*; zone 1, 2, 3.
Highbush Blueberry, *V. corymbosum*; zone 1, 2.
Greentwig Huckleberry, *V. elliottii*; zone 1, 2.
Rabbiteye Huckleberry, *V. ashei*; zone 1, 2, 3.
Running Huckleberry, *V. crassifolium*; evergreen; zone 1, 2.

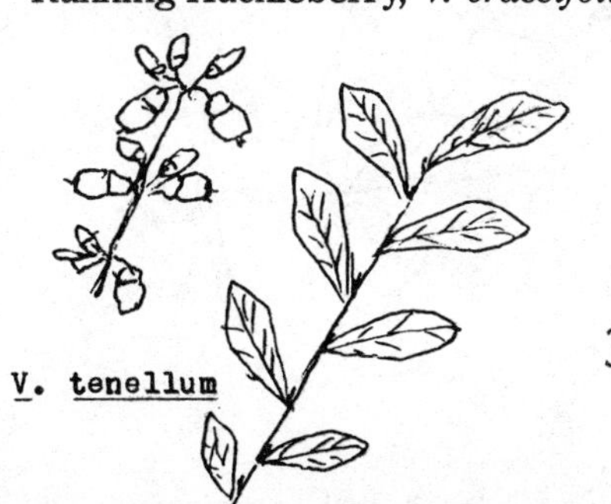

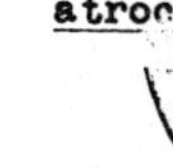

Boston-Ivy, Virginia Creeper

Parthenocissus spp.

Grape Family

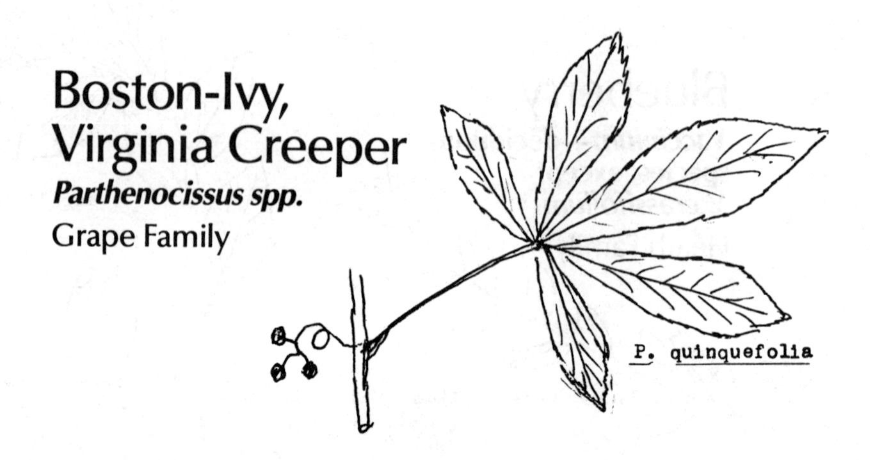

GROWTH HABIT: Deciduous woody vines, high-climbing by disk-bearing tendrils and/or aerial roots
LEAVES: Simple or compound; see below
FLOWERS: Small greenish and in clusters borne terminally or opposite the leaves
FRUITS: 1/4″ wide bluish-black berries
SEEDS: 1-4
LIGHT: Partial or full sun
SOIL: Tolerant
MOISTURE: Moist
SEASONAL ASPECT: Brightly colored leaves in autumn
ZONE: 1, 2, 3
USES: Arbor, trellis, wall, masonry, tree
CARE: None
PROPAGATION BY: Seedlings and cuttings
NATIVE OF: See below
OTHER: Boston Ivy, *P. tricuspidata*; Japan and China; Japanese or Boston Ivy; leaves 3-lobed or 3-foliate; several cultivated varieties.
Virginia Creeper, *P. quinquefolia*; e. United States; leaves 5-foliate; several cultivated varieties.

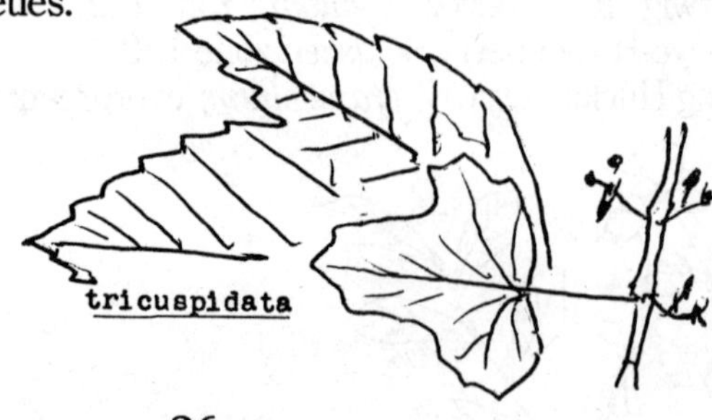

Bottle-Brush

Callistemon speciosus

Myrtle Family

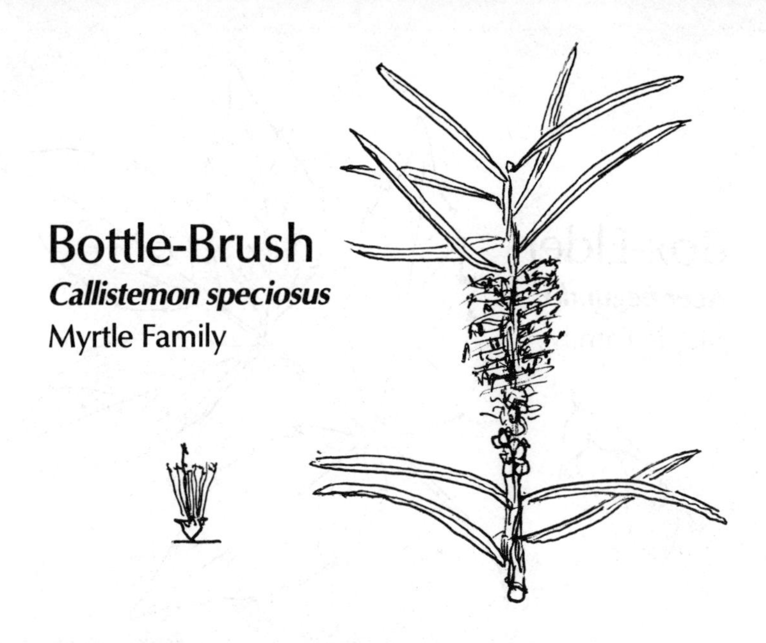

GROWTH HABIT: Rather large deciduous shrub
LEAVES: Narrow, to 3″ long and entire
FLOWERS: In dense spikes toward twig tips, small but with 1″ long filaments bearing yellow anthers
FRUITS: Small globular capsule
SEEDS: Tiny
LIGHT: Full sun
SOIL: Most any fertile type
MOISTURE: Moist to dry
SEASONAL ASPECT: Summer flowers
ZONE: 1, 2
USES: Occasional, specimen
CARE: Perhaps some pruning
PROPAGATION BY: Cuttings
NATIVE OF: Australia
OTHER: One or more additional species are cultivated. They have dark colored anthers.

Box-Elder

Acer negundo

Maple Family

GROWTH HABIT: Large, spreading deciduous tree with closely furrowed corky bark and shining green twigs
LEAVES: Oppostie, pinnate, with 3-5 (7) leaflets, each usually toothed or lobed
FLOWERS: Small, yellowish-green and appearing before the leaves in a pendulous arrangement
FRUITS: Paired and winged nutlets
SEEDS: 1 per nutlet
LIGHT: Sun or partial shade
SOIL: Silty loam
MOISTURE: Moist to wet
SEASONAL ASPECT: Yellow leaves in autumn
ZONE: (1), 2, 3
USES: As any other tree
CARE: None
PROPAGATION BY: Seedlings
NATIVE OF: e. North America

African Boxwood

Myrsine africana
Myrsine Family

GROWTH HABIT: Evergreen shrub to 4′ high
LEAVES: Alternate, broadly elliptic and leathery
FLOWERS: Very small, axillary and 3-8 together, 4-parted and unisexual
FRUITS: Purplish and berry-like
SEEDS: 1
LIGHT: Partial shade
SOIL: Fertile sandy loam
MOISTURE: Dry
SEASONAL ASPECT: None
ZONE: 1 but only where protected from extreme cold
USES: Specimen
CARE: None
PROPAGATION BY: Seedlings and cuttings
NATIVE OF: Africa

Boxwood

Buxus spp.
Box Family

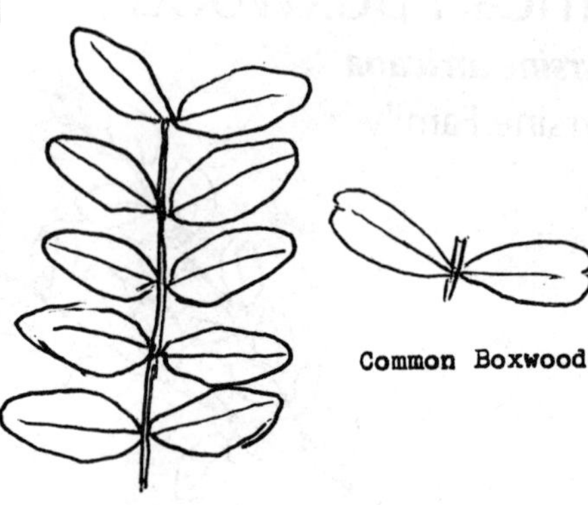

GROWTH HABIT: Compact, evergreen, shallow-rooted shrubs
LEAVES: Opposite, leathery, entire, small, short petioled and shining dark-green above
FLOWERS: Male and female separate on same plant, both small, without petals and axillary
FRUITS: Small, sessile, 3-pointed capsule
SEEDS: Shining black
LIGHT: Full or partial sun
SOIL: Rich loamy type
MOISTURE: Moist but well drained
SEASONAL ASPECT: None
ZONE: 1, 2, 3
USES: Edging, hedge, unit arrangement
CARE: Almost none, except to keep dogs away
PROPAGATION BY: Cuttings
NATIVE OF: Europe, n. Africa, w. Asia
OTHER: Common Boxwood, *B. sempervireus*; larger and more spreading shrub.
English Boxwood, *B. microphylla*; very compact with small leaves and slow growth. Several varieties of each are cultivated.

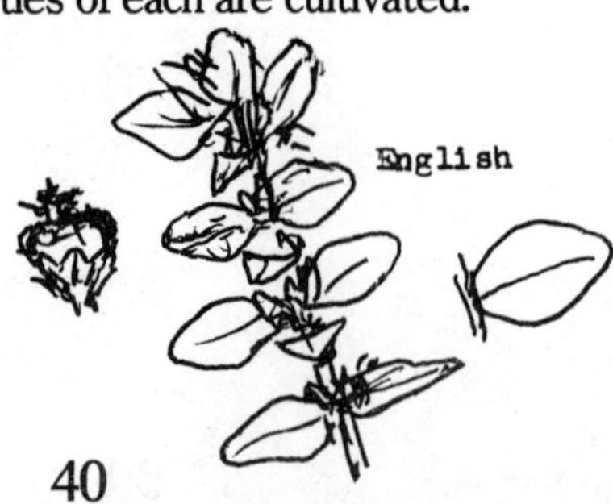

Buckeye

Aesculus spp.

Horse-Chestnut Family

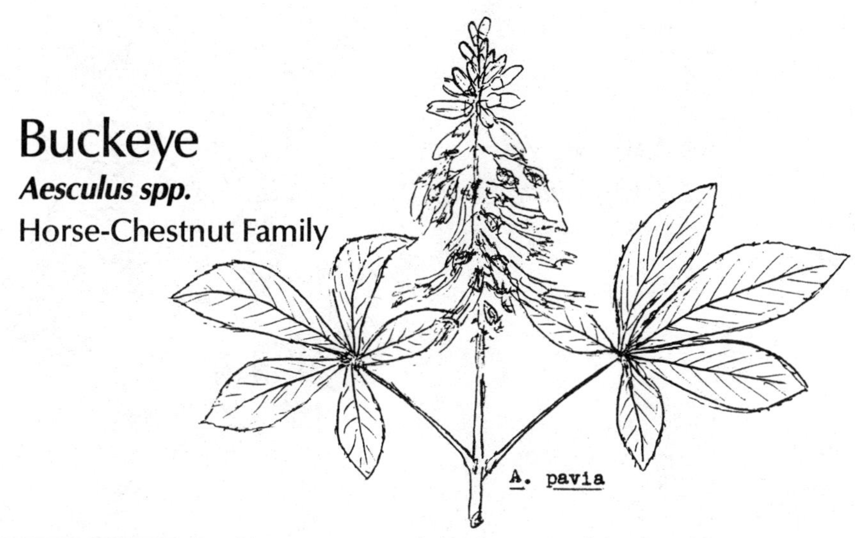

GROWTH HABIT: Deciduous trees and shrubs; one of the first plants to leaf-out in the spring
LEAVES: Opposite, digitately 5-9 foliate, leaflets to 8″ long, oblanceolate and toothed
FLOWERS: Irregular, about 1″ long and in showy terminal panicles
FRUITS: 1-2″ wide leathery capsule
SEEDS: 1-3, lusterous brown and to 1″ wide
LIGHT: Half shade
SOIL: Rich alluvial type
MOISTURE: Moist
SEASONAL ASPECT: Flowering time soon after leaves are grown
ZONE: 1, 2, 3
USES: Wet or moist bank, freestanding, specimen
CARE: None
PROPAGATION BY: Seedlings
NATIVE OF: e. United States
OTHER: Red Flowered Buckeye, *A. pavia*; shrub with red flowers. Green Flowered Buckeye; *A. sylvatica*; shrub with yellow-green flowers. Ohio Buckeye, *A. octandra*; tree. Foliage and fruits of all are poisonous.

Buckleya

Buckleya distichophylla

Sandalwood Family

GROWTH HABIT: Deciduous shrub to 8′ high
LEAVES: Opposite, borne in 2 ranks and sessile
FLOWERS: Small, greenish, male and female on separate plants, no petals but with relatively large bracts
FRUITS: An almost inch long elongated drupe
SEEDS: 1
LIGHT: Partial shade
SOIL: See below
MOISTURE: See below
SEASONAL ASPECT: None
ZONE: 3
USES: Specimen
CARE: None
PROPAGATION BY: Seeds
NATIVE OF: s. Appalachians
OTHER: A very rare shrub believed to be partially parasitic on the roots of Hemlock *(Tsuga canadensis)* and should be planted near one. It does the Hemlock no apparent harm. This plant needs the protection of cultivation.

Buckthorn

Bumelia lycioides

Sapodilla Family

GROWTH HABIT: Semievergreen shrub to 12′ high, often spiny

LEAVES: Alternate or clustered on spur shoots, elliptic to oblanceolate, entire and rather thin

FLOWERS: Small, white or whitish, and in many-flowered clusters along twigs and spur twigs

FRUITS: Drupe-like, almost 1/2″ long and longer than wide

SEEDS: Few

LIGHT: Partial shade

SOIL: Rich loam or alluvium

MOISTURE: Moist to wet

SEASONAL ASPECT: Flowers appear after leaves are grown

ZONE: 1, 2

USES: Interest

CARE: None

PROPAGATION BY: Seedlings or cuttings

NATIVE OF: e. United States

OTHER: *B. tenax* is a lower coastal plain form differing in that the leaves are very silky-hairy beneath.

Buffalo-Nut, Oil-Nut

Pyrularia pubera

Sandalwood Family

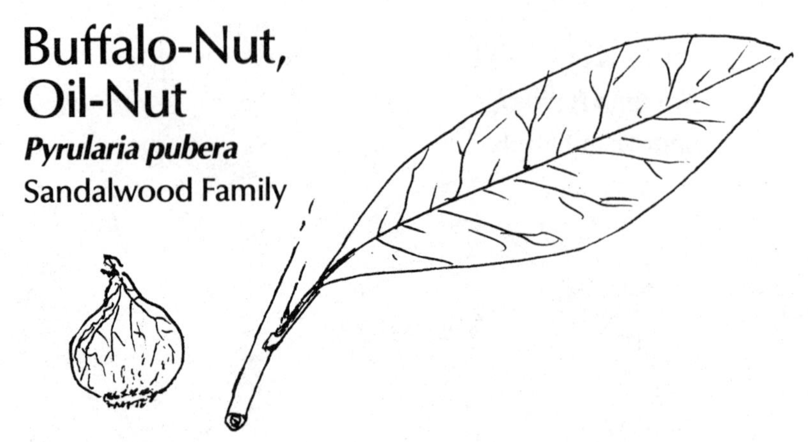

GROWTH HABIT: Rather weak, deciduous shrub to 10′ high
LEAVES: Alternate, thin, widely elliptic, entire, veiny and to 5″ long
FLOWERS: Male and female on separate plants, or as separate flowers on the same plant; small, whitish and in short spikes
FRUITS: 1″ wide, green
SEEDS: 1
LIGHT: Half shade or more
SOIL: Rich
MOISTURE: Moist
SEASONAL ASPECT: None
ZONE: 2, 3
USES: In hardwood area to help create naturalistic effect
CARE: None
PROPAGATION BY: Seeds
NATIVE OF: s.e. United States
OTHER: May be a root parasite, dependent on hardwood species as Oak, Hickory or other woody plants, producing no apparent harm; fruits toxic.

Bugle-Weed

Ajuga reptans

Mint Family

GROWTH HABIT: Prostrate, stoloniferous evergreen; glabrous or nearly so; mat forming
LEAVES: Opposite, elliptic to obovate with entire or wavy margins
FLOWERS: Violet-blue, 1/2″ long and in terminal spikes
FRUITS: 4 nutlets per flower
SEEDS: Partial or full shade
LIGHT: Most types
SOIL: Moist
MOISTURE: Late spring flowers
SEASONAL ASPECT: 1, 2, 3
ZONE: Ground cover, bank, rockery, border
USES: None
CARE: Stolons
PROPAGATION BY: Europe
NATIVE OF: Varieties with purplish, bronzed and variegated leaves are grown.

Bush-Aster

Olearia haastii

Composite Family

GROWTH HABIT: Much branched, evergreen shrub to 6′ high
LEAVES: Alternate, crowded, ovate, entire and to 1″ long; whitened beneath
FLOWERS: Rays white, center yellow, 1/2″ wide
FRUITS: Achene
SEEDS: 1
LIGHT: Full sun
SOIL: Any well drained type
MOISTURE: Dry to moist
SEASONAL ASPECT: None
ZONE: 1, 2
USES: Border, bank, specimen
CARE: Some pruning in early spring
PROPAGATION BY: Seedlings
NATIVE OF: New Zealand

Bush Cinquefoil

Potentilla arbuscula

Rose Family

GROWTH HABIT: Deciduous shrub to 3′ high

LEAVES: Palmate with mostly five leaflets, each 1″ long or less, gray-green, silky and numerous

FLOWERS: Bright yellow; 5-parted and about 1″ wide

FRUITS: Small achenes

SEEDS: 1 per achene

LIGHT: Full sun

SOIL: Almost any type

MOISTURE: Dry to moist

SEASONAL ASPECT: An abundance of showy flowers in late spring

ZONE: 1, 2, 3

USES: Foundation, front for shrub border, edging

CARE: None

PROPAGATION BY: Late spring, early summer cuttings

NATIVE OF: Himalayas

OTHER: *P. fruticosa*; world wide; many named varieties; some white.

Bush Clover, Bladder Senna

Colutea arborescens

Legume Family

GROWTH HABIT: Fast growing, deciduous shrub to 8′ high
LEAVES: Alternate, pinnate with 7-11 leaflets each to 1″ long
FLOWERS: Yellow, fragrant, pear shaped and nearly 1″ wide
FRUITS: Much inflated pods to 2″ long
SEEDS: Several
LIGHT: Full sun
SOIL: Any fertile type
MOISTURE: Well drained
SEASONAL ASPECT: Showy flowers after leaves are grown and inflated bronze pods in late summer
ZONE: 1, 2, 3
USES: Screen, shrub, border
CARE: Prune in early spring to produce bushy plant
PROPAGATION BY: Seeds, summer cutting
NATIVE OF: Mediterranean region

Bush Honeysuckle

Diervilla sessilifolia

Honeysuckle Family

GROWTH HABIT: Deciduous shrub to 4′, somewhat stoloniferous
LEAVES: Opposite, broadly lanceolate, long-tapering, entire and to 5″ long
FLOWERS: Yellow, tubular with 5 lobes, in terminal clusters
FRUITS: 1/2″ long capsule
SEEDS: Very small
LIGHT: Shifting shade
SOIL: Any clay loam
MOISTURE: Moist
SEASONAL ASPECT: Early summer flowers
ZONE: 3
USES: Bank, shrub cover,
CARE: None
PROPAGATION BY: Cuttings
NATIVE OF: s.e. United States

Bush Poppy

Dendromecon rigida

Poppy Family

GROWTH HABIT: Rather weakly erect evergreen shrub from 2-10′ high
LEAVES: Whitened, leathery, usually narrow and to 4″ long
FLOWERS: Bright yellow, to 2″ wide and 4-parted
FRUITS: Capsule, elongated, curved and grooved; to 3″ long
SEEDS: Numerous
LIGHT: Full sun
SOIL: Fertile loam
MOISTURE: Dry
SEASONAL ASPECT: Flowers in very early spring
ZONE: 1
USES: Against wall, facing
CARE: Some trimming
PROPAGATION BY: Seeds
NATIVE OF: California

Butchers Broom

Ruscus aculeatus

Lily Family

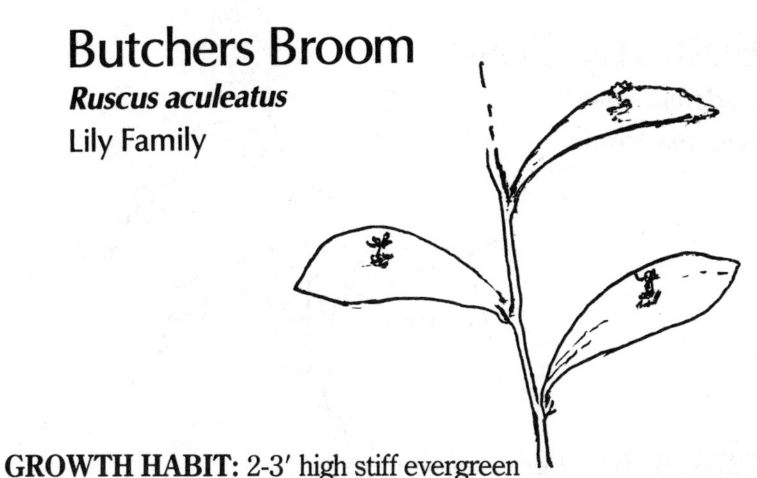

GROWTH HABIT: 2-3′ high stiff evergreen
LEAVES: The ovate, spine-tipped leaf-like structures are actually stem sections; the leaves are the minute non-green scales below.
FLOWERS: Male and female on separate plants; both small, greenish and attached to the spine-tipped stem sections
FRUITS: 1/2″ wide, red or yellow berries
SEEDS: Few
LIGHT: Partial or full sun
SOIL: Fertile, sandy loam
MOISTURE: Moist
SEASONAL ASPECT: None
ZONE: 1, 2
USES: Interest or specimen
CARE: None
PROPAGATION BY: Seeds and seedlings
NATIVE OF: Madeira Islands
OTHER: Cuttings are useful in dried arrangements and may be dyed.

Butterfly-Bush

Buddleja davidii

Logania Family

GROWTH HABIT: Deciduous shrub to 10′ high

LEAVES: Opposite, lanceolate to broadly so, long pointed, finely toothed and densely white-hairy beneath

FLOWERS: Lilac with orange centers, small but numerous in elongate terminal inflorescence

FRUITS: Small capsule

SEEDS: Few

LIGHT: Full or partial sun

SOIL: Fertile or poor sandy types

MOISTURE: Moist

SEASONAL ASPECT: Flowers in mid or late summer

ZONE: 1, 2, 3

USES: Occasional flowering shrub, specimen

CARE: Some pruning

PROPAGATION BY: Cuttings

NATIVE OF: China

OTHER: A number of named varieties of this species are grown. Fountain Buddleja; *B. alternifolia*; China; A vigorous grower with widely arching branches to 12′ wide, blooms abundantly in mid-spring.

Butterfly-Weed, Pleurisy-Root

Asclepias tuberosa

Milkweed Family

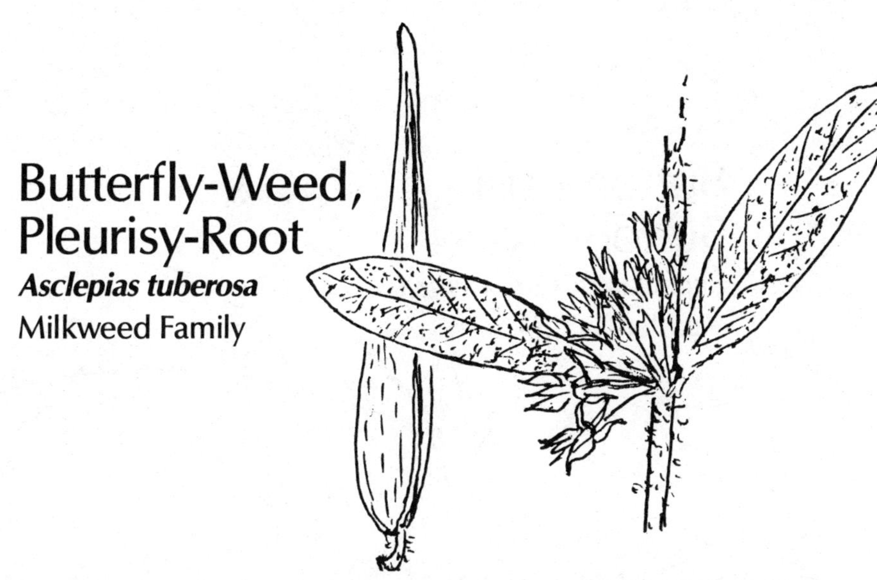

GROWTH HABIT: Deciduous perennial herb to 2′ high from strong roots
LEAVES: Alternate, at least below, entire, pubescent and to 2 1/2″ long
FLOWERS: Orange-red to yellow, in terminal and axillary clusters; 5-parted
FRUITS: Pod to 4″ long
SEEDS: Flat, black and with tuft of silky hairs
LIGHT: Full sun
SOIL: Sandy or clay loam
MOISTURE: Dry to moist
SEASONAL ASPECT: Showy flowers, late spring and summer
ZONE: 1, 2, 3
USES: Interest, single planting
CARE: None
PROPAGATION BY: Seeds and transplants
NATIVE OF: s.e. United States
OTHER: A hardy and attractive plant once established will be a joy for years.

Button-Bush, Button-Wood

Cephalanthus occidentalis

Madder Family

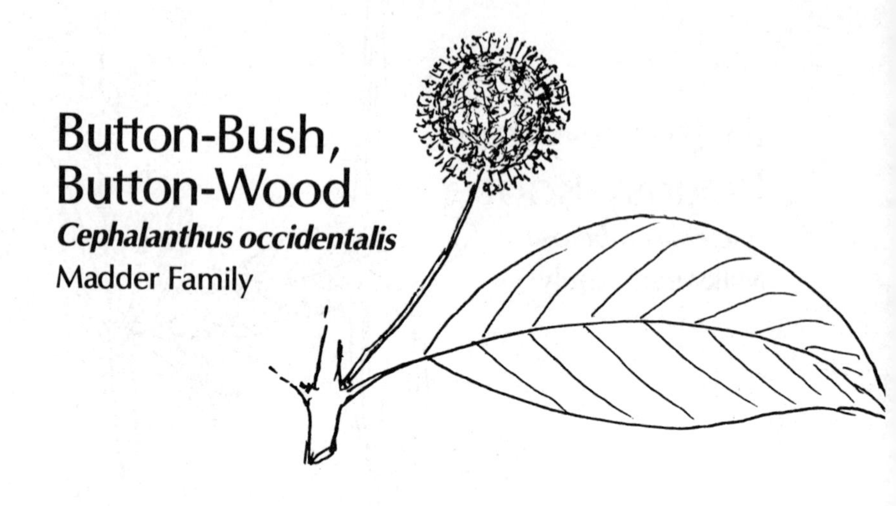

GROWTH HABIT: Deciduous shrub
LEAVES: Opposite or whorled, elliptic, long pointed and entire
FLOWERS: White, small very numerous and crowded into spherical heads 1″ or more in diameter
FRUITS: Small angular nutlets
SEEDS: 1 per nutlet
LIGHT: Sun or partial shade
SOIL: Most any type, alluvium best
MOISTURE: Wet
SEASONAL ASPECT: Summer flowers
ZONE: 1, 2, 3
USES: Along wet margins
CARE: None
PROPAGATION BY: Seedlings
NATIVE OF: e. North America

California Incense Cedar

Calocedrus decurrens

Cypress Family

GROWTH HABIT: Columnar evergreen tree with reddish-brown scaly bark
LEAVES: Opposite, lustrous dark green, scale-like and decurrent
FLOWERS: Male and female separate, very small, greenish
FRUITS: 1″ long reddish-brown cones
SEEDS: 2 long-winged per cone scale
LIGHT: Lots of sun
SOIL: Tolerant to most well drained types
MOISTURE: Moist to dry
SEASONAL ASPECT: None
ZONE: 2, 3
USES: Specimen
CARE: None
PROPAGATION BY: Seedlings
NATIVE OF: w. United States
OTHER: Formerly *Libocedrus*

Camellia

Camellia japonica

Tea Family

GROWTH HABIT: Compact evergreen shrub

LEAVES: Alternate, oval to elliptic, shallowly toothed, dark green and glossy

FLOWERS: Axillary, very showy, white to red with color solid or variegated, petals numerous

FRUITS: 1″ leathery capsules

SEEDS: 1-3

LIGHT: Partial shade

SOIL: Fertile sandy loam

MOISTURE: Well drained but moist

SEASONAL ASPECT: Flowering time in early spring

ZONE: 1, 2

USES: Flowers, freestanding, unit plantings, hedges, borders

CARE: Pruning, fertilizing and guarding against scale

PROPAGATION BY: Cuttings, grafting, seeds

NATIVE OF: Japan

OTHER: Very popular and easily available in many varieties. In grafting, a scion is usually fixed to a *C. sasanqua* stock.

Sasanqua

Camellia sasanqua

Tea Family

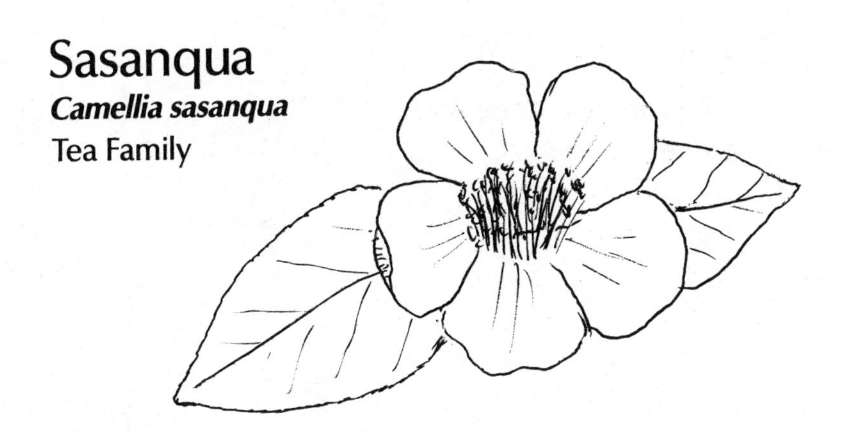

GROWTH HABIT: Compact evergreen shrub similar to *C. japonica*
LEAVES: Alternate, shallowly toothed, glossy above, 1 1/2 - 3″ long
FLOWERS: White to rose, 5 or more petals, slightly fragrant
FRUITS: Leathery capsule 1″ wide
SEEDS: 1-3
LIGHT: Some shade, excessive sun yellows the leaves
SOIL: Fertile sandy loam
MOISTURE: Moist but well drained
SEASONAL ASPECT: Flowering time in late fall
ZONE: 1, 2
USES: Foundations, hedges, accent planting
CARE: Prune, fertilize and control scale
PROPAGATION BY: Seeds, seedlings, cuttings
NATIVE OF: China, Japan
OTHER: This plant is commonly referred to by its specific name while *C. japonica* is referred to by its generic name.

Camphor-Tree

Cinnamomum camphora

Laurel Family

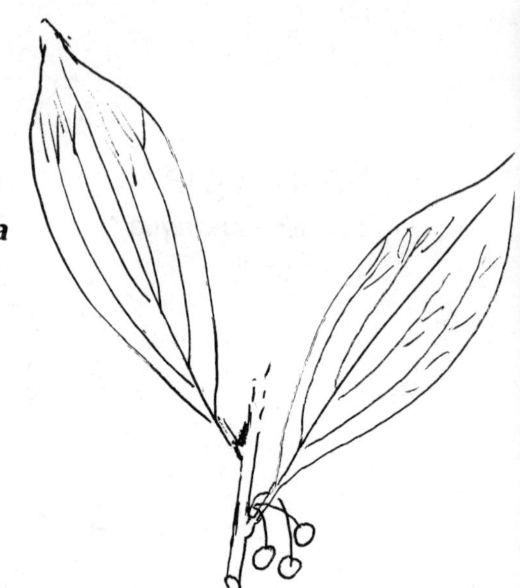

GROWTH HABIT: Small, semi-evergreen tree
LEAVES: Alternate, ovate, long pointed, aromatic, smooth
FLOWERS: Inconspicuous
FRUITS: Black berries about 1/4″ wide
SEEDS: Few per berry
LIGHT: Full sun
SOIL: Tolerant
MOISTURE: Moist to dry
SEASONAL ASPECT: Spring when new growth is very light green
ZONE: 1, 2
USES: Freestanding, specimen, interest
CARE: None
PROPAGATION BY: Seeds, seedlings
NATIVE OF: e. Asia
OTHER: Hardy in Zone 1 into North Carolina; however, extreme cold spells may kill it back to ground level but it suckers up in spring.

Candytuft

Iberis sempervirens

Mustard Family

GROWTH HABIT: Low evergreen perennial to 1′ high; flowering stems unbranched
LEAVES: To 1″ long, entire, glabrous, linear, blunt tipped
FLOWERS: White or tinted, terminal, 4-parted, outer two petals longer
FRUITS: 1/4″ long, 2 celled capsule
SEEDS: 2
LIGHT: Partial sun
SOIL: Any fertile porous type
MOISTURE: Moist to dry
SEASONAL ASPECT: Flowers in spring
ZONE: 1, 2, 3
USES: Edging
CARE: Weeding
PROPAGATION BY: Seeds
NATIVE OF: e. Europe
OTHER: *I. gibraltarica*; Spain; leaves mostly toothed; flowers reddish to white; flowering stems branched.

Cape Jasmine

Gardenia jasminoides

Madder Family

GROWTH HABIT: Evergreen shrub to 6′
LEAVES: Opposite, more or less elliptic, to 4″ long, entire and glossy green
FLOWERS: White, usually doubled, 2″ or more wide and very fragrant
FRUITS: Fleshy, 1″ long, ribbed and orange-colored
SEEDS: Several
LIGHT: Full sun for best growth and flowers
SOIL: Fertile sandy loam
MOISTURE: Moist to dry
SEASONAL ASPECT: Summer flowers and fragrance
ZONE: 1, 2, 3
USES: Occasional shrub, border
CARE: Transplant carefully
PROPAGATION BY: Cuttings
NATIVE OF: China, but first throught to be from Cape of Good Hope.
OTHER: Several quite different varieties are grown.

Carolina Jessamine

Gelsemium sempervirens

Longania Family

GROWTH HABIT: Mostly evergreen woody vine climbing by twining to 15-20′
LEAVES: Opposite, glabrous, lanceolate and entire
FLOWERS: Bright yellow, to 1 1/2″ long with open centers and borne in leaf axils; fragrant
FRUITS: Flattened 1/2″ long capsules
SEEDS: Few
LIGHT: Full sun for flowers
SOIL: Tolerant
MOISTURE: Moist to dry
SEASONAL ASPECT: Flowers in early spring
ZONE: 1, 2
USES: Trellis, mailbox post, decoration, specimen
CARE: None
PROPAGATION BY: Seedlings or transplants
NATIVE OF: s.e. United States
OTHER: Poisonous

Cassia

Cassia bicapsularis

Legume Family

GROWTH HABIT: Tardily deciduous shrub to 6′ high
LEAVES: Alternate, even-pinnate with 3-5 pairs of oblong, rounded-tip leaflets
FLOWERS: Yellow, 1/2″ long and in axillary racemes
FRUITS: 2 1/2 - 5″ long pods
SEEDS: Several, shining brown
LIGHT: Three-fourths sun
SOIL: Tolerant
MOISTURE: Moist
SEASONAL ASPECT: Late summer flowers
ZONE: 1, 2
USES: Specimen, occasional shrub, patio
CARE: None
PROPAGATION BY: Seeds
NATIVE OF: Tropics
OTHER: *C. corymbosa*; Argentina; similar to above but with only 2 - 3 pairs of leaflets and each leaflet with a pointed tip.

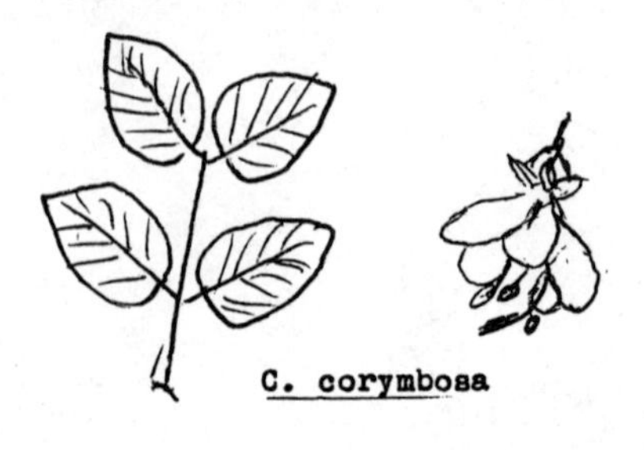

Cast Iron Plant

Aspidistra elatior

Lily Family

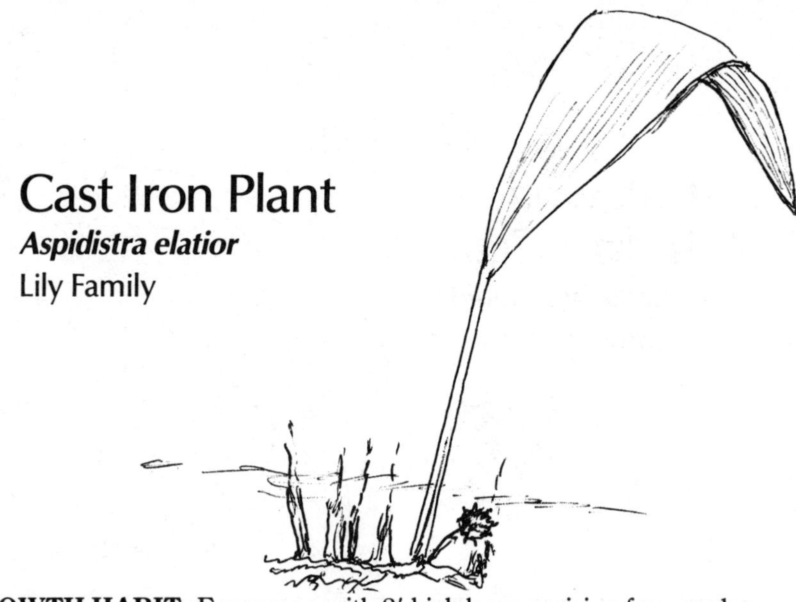

GROWTH HABIT: Evergreen with 2′ high leaves arising from underground matted rhizomes
LEAVES: 3-4″ wide, lanceolate and smooth
FLOWERS: Brownish purple, 1″ wide at ground level or below in January
FRUITS: Small berry
SEEDS: 1 per berry
LIGHT: Full or partial shade
SOIL: Fertile loam
MOISTURE: Moist
SEASONAL ASPECT: None
ZONE: 1, 2
USES: Interest or accent
CARE: None
PROPAGATION BY: Clump division
NATIVE OF: Himalayas, China and Japan
OTHER: Some forms have variegated foliage.

Catawba, Indian Cigar

Catalpa bignonioides

Trumpet-Creeper Family

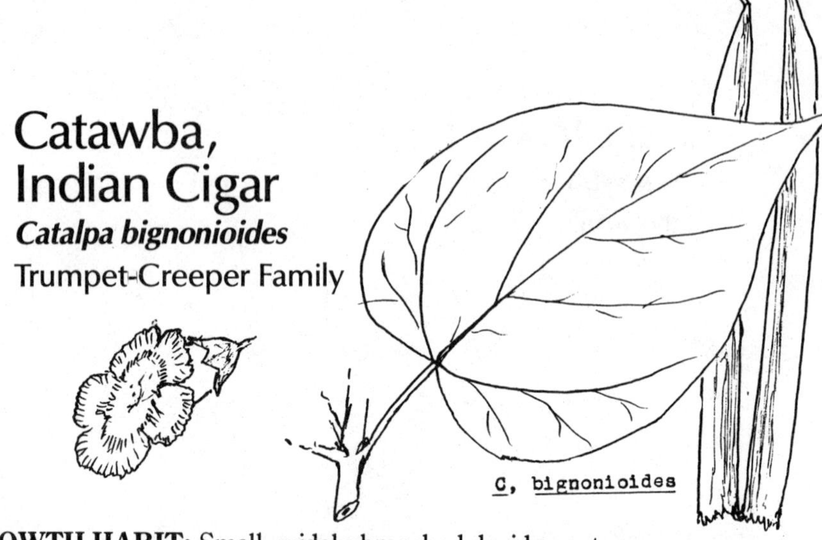

GROWTH HABIT: Small, widely-branched deciduous tree

LEAVES: Opposite or whorled, broadly heart-shaped, entire or occasionally with 2 shallow lobes to 8″ long

FLOWERS: White with yellow strips and purple spots, about 2″ wide with open center

FRUITS: 1′ long pod of pencil size

SEEDS: Winged and numerous

LIGHT: Full sun

SOIL: Tolerant to most types

MOISTURE: Moist

SEASONAL ASPECT: Early summer flowers

ZONE: 1, 2, 3

USES: Freestanding, background

CARE: None

PROPAGATION BY: Seeds or seedlings

NATIVE OF: s.e. United States

OTHER: Sometimes grown by fisherman for the sphinx moth caterpillars that feed on the leaves. *C. speciosa*; s.e. United States; has slightly larger flowers.

Cat's Claw-Vine

Macfadyena unguis-cati
(Doxantha, Bignonia violacea)

Trumpet Creeper Family

GROWTH HABIT: Evergreen shrubby vine, climbing by means of 3-branched claw-like tendrils
LEAVES: Opposite, compound with 2 leaflets and a terminal 3-parted claw-like tendril; leaflets lanceolate; entire and to 2″ long
FLOWERS: Large, showy, yellow with orange markings, 2″ long and as wide or wider, center open
FRUITS: 1′ long pod
SEEDS: Winged
LIGHT: Full or partial sun
SOIL: Most any fertile type
MOISTURE: Moist
SEASONAL ASPECT: Early spring flowers
ZONE: 1
USES: Wood or masonry walls
CARE: None
PROPAGATION BY: Seeds
NATIVE OF: w. Indies

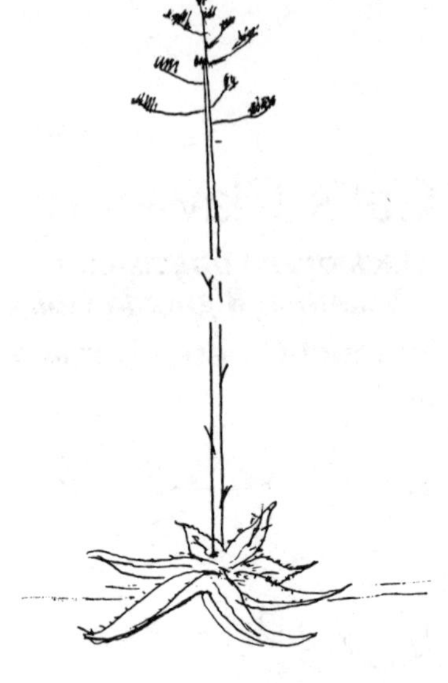

Century Plant

Agave americana

Amaryllis Family

GROWTH HABIT: Extremely robust evergreen forming a basal rosette 6 to 8′ wide

LEAVES: Large, thick, tough, spiny-margined 3-8″ wide and to 3′ long, arising from underground stem

FLOWERS: 2″ long, white, numerous, erect and raised on 10-20′ high flower stalk

FRUITS: Capsule

SEEDS: Numerous

LIGHT: Full sun

SOIL: Fertile, sandy

MOISTURE: Dry

SEASONAL ASPECT: None

ZONE: 1, 2

USES: Open corner, tub plant for patio or terrace, interest

CARE: None

PROPAGATION BY: Suckers

NATIVE OF: Tropical America

OTHER: Flowering occurs only every several years following which plant dies but may leave suckers. Some varieties have striped or margined leaves.

Ceratiola

Ceratiola ericoides

Crowberry Family

GROWTH HABIT: Much-branched evergreen shrub to 6′ high
LEAVES: Small, numerous, linear, 1/2″ long and borne in 6 ranks
FLOWERS: Small, male and female on separate plants
FRUITS: Small, drupe-like and olive in color
SEEDS: 2
LIGHT: Full sun
SOIL: Sand or sandy loam
MOISTURE: Dry
SEASONAL ASPECT: None
ZONE: 1, 2
USES: Occasional shrub
CARE: None
PROPAGATION BY: Seeds
NATIVE OF: s.e. United States
OTHER: Very difficult to transplant.

Chaste-Tree

Vitex agnus-castus

Vervain Family

GROWTH HABIT: Large, deciduous shrub with densely hairy twigs; to 10′ high
LEAVES: Opposite, palmate with 5-7 leaflets, leaflets lanceolate, 2-4″ long, few teeth, if any; dark green above, grayish beneath
FLOWERS: Small, pale blue and in dense, mostly terminal spikes to 6″ long
FRUITS: Small, dark, hard, dry structure
SEEDS: 1-4
LIGHT: Lots of sun
SOIL: Most any type
MOISTURE: Moist to dry
SEASONAL ASPECT: Flowers in late summer and autumn
ZONE: 1, 2
USES: Background
CARE: None
PROPAGATION BY: Seeds and seedlings
NATIVE OF: s. Europe
OTHER: Several varieties are cultivated.

Cherry-Laurel

Prunus caroliniana, and other evergreen species

Rose Family

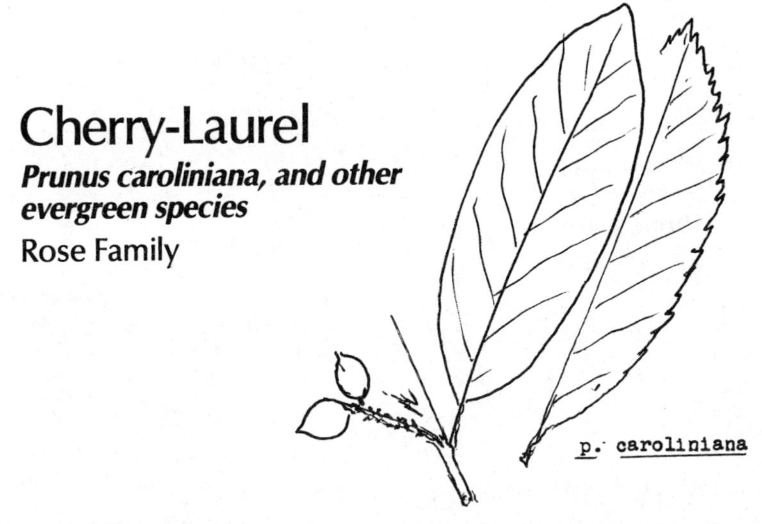

GROWTH HABIT: Small to medium evergreen trees frequently pruned and kept as shrubs
LEAVES: Alternate, glabrous, elliptic, to 3 1/2″ long and entire or finely or coarsely toothed
FLOWERS: Small, white and numerous on short racemes
FRUITS: Blue, thin-pulped drupe 1/2″ long
SEEDS: 1 and relatively large
LIGHT: Sun or partial shade
SOIL: Tolerant
MOISTURE: Moist
SEASONAL ASPECT: None
ZONE: 1, 2
USES: As shrub or small tree
CARE: Some pruning if shrub form desired
PROPAGATION BY: Seedlings
NATIVE OF: s.e. United States
OTHER: *P. laurocerasus*; European Cherry-Laurel; similar and in several varietal forms.
P. lusitanica; Portugal-Laurel; also similar and in several varietal forms.

Flowering Cherry

Prunus spp.

Rose Family

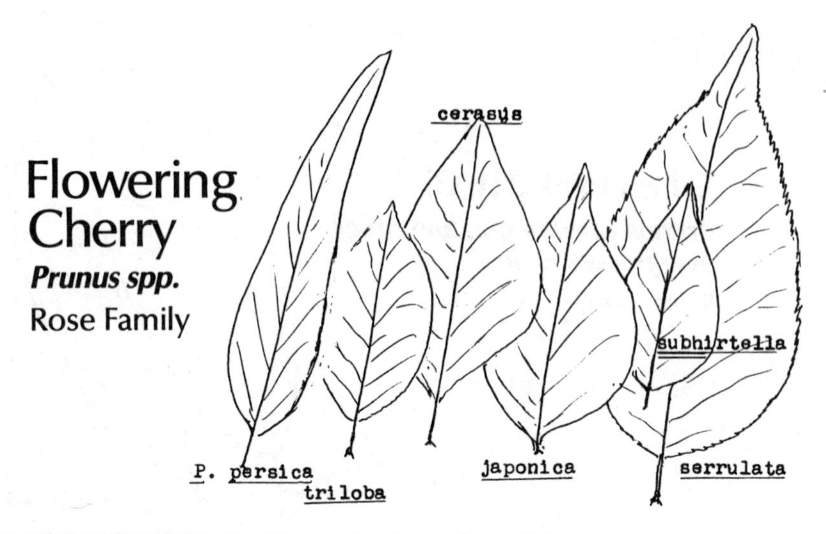

GROWTH HABIT: Deciduous shrubs and small trees
LEAVES: Alternate, lanceolate to ovate, obscurely, finely or sharply toothed and to 3″ long
FLOWERS: Profusely produced, white to pink, single or double
FRUITS: When produced, mostly red and to 1/2″ long
SEEDS: 1 per fruit
LIGHT: Full sun for best blossoms
SOIL: Fertile
MOISTURE: Moist
SEASONAL ASPECT: Early spring flowers just before the leaves
ZONE: 1, 2, 3
USES: Decoration, freestanding, street
CARE: None
PROPAGATION BY: Cuttings
NATIVE OF: Orient
OTHER: Flowering Peach; a variety of *P. persica*; China; deciduous tree-like shrub.

Flowering Almond; *P. triloba*; China; deciduous tree-like shrub with pink flowers.

Sour, Red or Pie Cherry; *P. cerasus*; Eurasia; small deciduous tree grown for flowers and fruits.

Dwarf Flowering Cherry; *P. japonica*; deciduous shrub to 4′ high; sometimes offered as Flowering Almond or *P. sinensis*.

Rosebud Cherry; *P. subhirtella*; Japan; deciduous, large shrub or small tree.

Japanese Flowering Cherry; *P. serrulata*; China, Japan and Korea; medium size deciduous tree.

Japanese Flowering Cherry; *P. sieboldii*; Japan; much like above but double-flowered. Each of the above is available in different varieties.

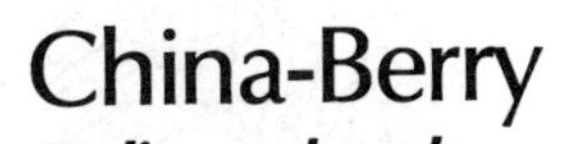

China-Berry

Melia azedarach

Mahogany Family

GROWTH HABIT: Deciduous tree in large and small varieties
LEAVES: Bipinnate, to 30″ long with numerous leaflets each ovate, toothed or lobed and to 2″ long
FLOWERS: Purplish, fragrant and in large open panicles
FRUITS: 1/2″ wide yellowish, somewhat fleshy drupes
SEEDS: 1 with fluted or ridged sides
LIGHT: Full or lots of sun
SOIL: Tolerant
MOISTURE: Moist to dry
SEASONAL ASPECT: None
ZONE: 1, 2, 3
USES: Background, shade, freestanding
CARE: None
PROPAGATION BY: Seedlings
NATIVE OF: China
OTHER: Cultivated in several different forms; fruit pulp toxic.

China Fir

Cunninghamia lanceolata

Taxodium Family

GROWTH HABIT: Narrow evergreen tree suckering freely at base
LEAVES: Narrow, long tapering, sharp pointed, stiff with two white bands beneath and appearing two-ranked
FLOWERS: Male as clustered, yellowish catkins; female as 1 1/2″ long cones
FRUITS: Leathery, brownish cones
SEEDS: Narrowly winged
LIGHT: Partial or full sun
SOIL: Tolerant
MOISTURE: Dry to moist
SEASONAL ASPECT: None
ZONE: 1, 2, 3
USES: Freestanding
CARE: Little or none
PROPAGATION BY: Seedlings or root sprouts
NATIVE OF: China

Chinese Parasol-Tree

Firmiana simplex

Sterculia Family

GROWTH HABIT: Smooth barked, deciduous tree to 30′
LEAVES: Large, to 8″ wide, palmately 5-7 lobed, lobes pointed
FLOWERS: Small, greenish and in 1′ long panicles
FRUITS: To 3″ long and opening long before maturity into green leaf-like structures bearing wrinkled seeds along their margins
SEEDS: 1-3 per carpel
LIGHT: Full sun
SOIL: Tolerant
MOISTURE: Moist
SEASONAL ASPECT: None
ZONE: 1, 2
USES: Interest, freestanding, background, patio
CARE: None
PROPAGATION BY: Seedlings
NATIVE OF: China, Japan

Chinquapin

Castanea pumila

Beech Family

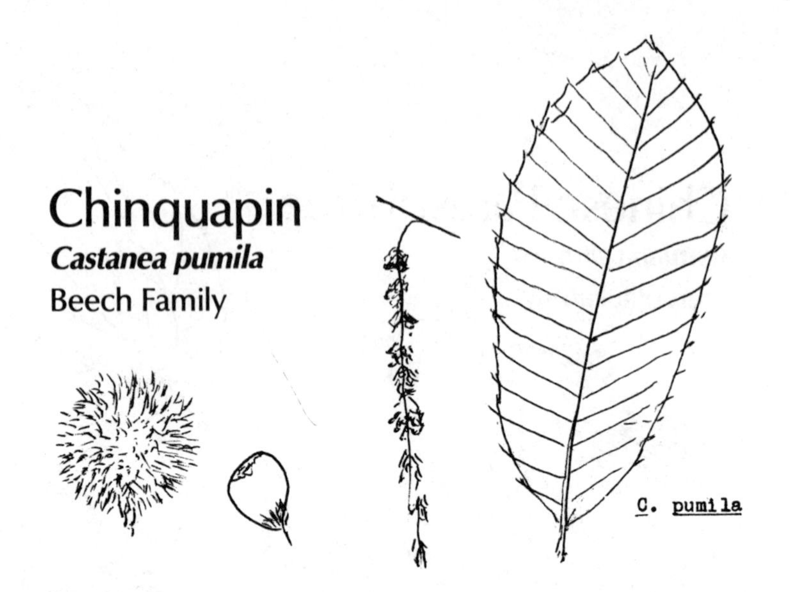

GROWTH HABIT: Uncommon deciduous shrub to 8′ high

LEAVES: Alternate, elliptic, 3-8″ long, usually pale and pubescent beneath, margin with bristle-tipped teeth

FLOWERS: Male in slender stiff catkins; female small and usually near base of male catkins

FRUITS: 1/2″ long thin-shelled nut inside spiny bur that splits 2-4 ways

SEEDS: Delightfully edible

LIGHT: Some shade

SOIL: Fertile loam

MOISTURE: Moist but well drained

SEASONAL ASPECT: None

ZONE: 1, 2, 3

USES: Interest

CARE: None

PROPAGATION BY: Seedlings

NATIVE OF: s.e. United States

OTHER: *C. alnifolia* of the coastal plain is a smaller and colony-forming plant.

Chokeberry

Aronia spp.

Rose Family

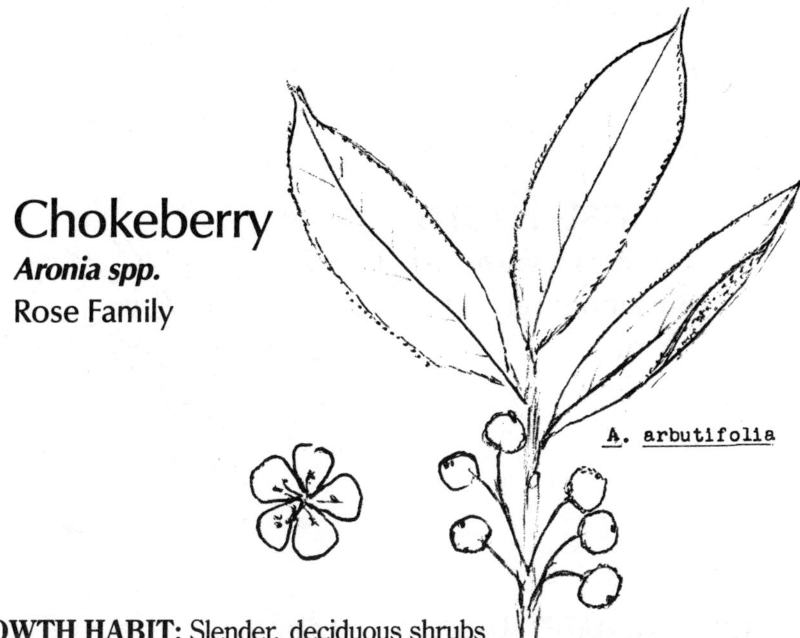

GROWTH HABIT: Slender, deciduous shrubs
LEAVES: Alternate, elliptic, serrate to 3″ long, and hairy or smooth beneath
FLOWERS: White, 1/2″ wide and in small clusters
FRUITS: Small, red or black and berry-like
SEEDS: Several
LIGHT: Sun or shade
SOIL: Rich in humus type
MOISTURE: Moist to wet
SEASONAL ASPECT: Spring flowers
ZONE: 1, 2, 3
USES: Occasional flowering shrubs for moist or wet locations
CARE: None
PROPAGATION BY: Seedlings or cuttings
NATIVE OF: e. North America
OTHER: *A. arbutifolia*; red fruits. *A. melanocarpa*; black fruits.

Christ-Thorn

Paliurus spina-christi

Buckthorn Family

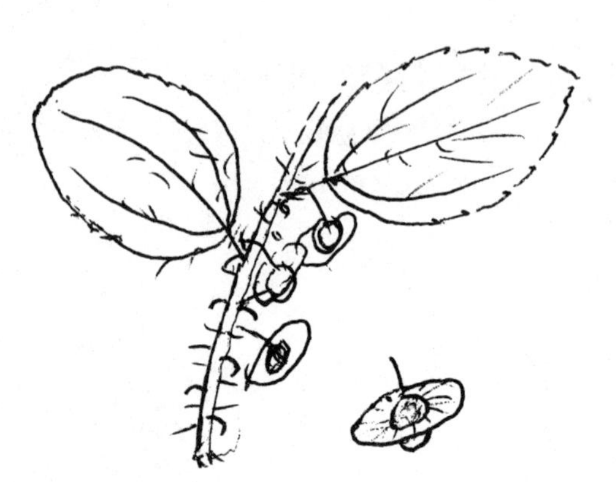

GROWTH HABIT: Deciduous spreading shrub or small tree
LEAVES: Alternate, ovate, to 1 1/2″ long, 3-veined, toothed with straight and hooked spines at leaf bases
FLOWERS: Small, greenish-yellow, 5-parted and mostly axillary
FRUITS: Dry, globose with horizontal wing, 3/4″ wide
SEEDS: 1-few
LIGHT: Plenty of sun
SOIL: Most any type
MOISTURE: Dry
SEASONAL ASPECT: None
ZONE: 1, 2, (3)
USES: Specimen
CARE: Barrier
PROPAGATION BY: Seeds, cuttings
NATIVE OF: s. Europe - n. China
OTHER: Believed by some to have been used as "the crown of thorns", others hold that it may have been *Ziziphus* that was used.

Christmas-Berry, Toyon, California-Holly

Heteromeles arbutifolia

Rose Family

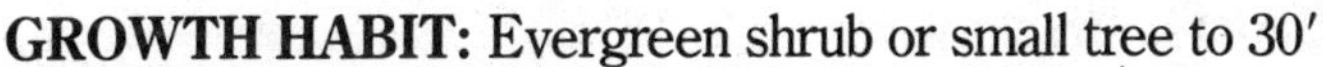

GROWTH HABIT: Evergreen shrub or small tree to 30′
LEAVES: Alternate, thick, oblong-lanceolate to 3″ long, sharply-serrate and shining above
FLOWERS: 1/4″ wide, white
FRUITS: 1/4″ long, bright red to yellow and persisting well into winter
SEEDS: Few
LIGHT: Full sun
SOIL: Fertile
MOISTURE: Dry
SEASONAL ASPECT: Red berries in winter
ZONE: 1, but only where it is protected from cold
USES: Terrace, patio, pot
CARE: Not very cold-hardy
PROPAGATION BY: Seedlings
NATIVE OF: s.w. United States
OTHER: see *Ardisia*.

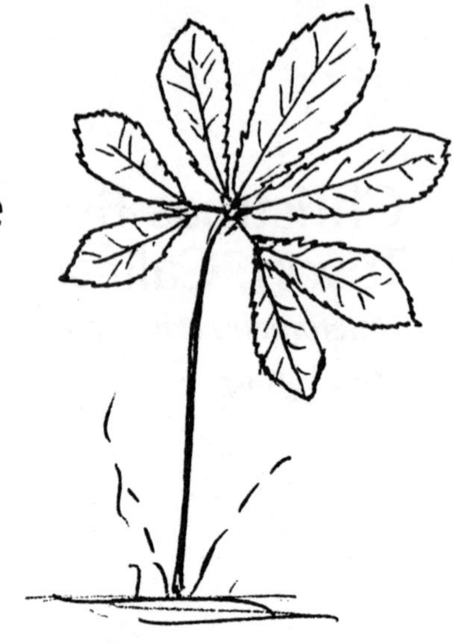

Christmas Rose

Helleborus niger

Buttercup Family

GROWTH HABIT: Erect evergreen to 18″ high, leaves and flowering stems arising from thick, fibrous roots
LEAVES: Supported by long petioles and deeply divided palmately into 3-7 divisions; thick and dark green
FLOWERS: Occurring in midwinter, cream to greenish and 2″ across; petals shorter than the stamens; sepals petal-like
FRUITS: 3-8 leathery pods per flower
SEEDS: Several per pod
LIGHT: Partial of full shade
SOIL: Fertile loam
MOISTURE: Moist
SEASONAL ASPECT: Mid-winter flowers
ZONE: 1, 2, 3
USES: Shady margin, border, wooded area
CARE: None
PROPAGATION BY: Seeds and seedlings
NATIVE OF: Europe

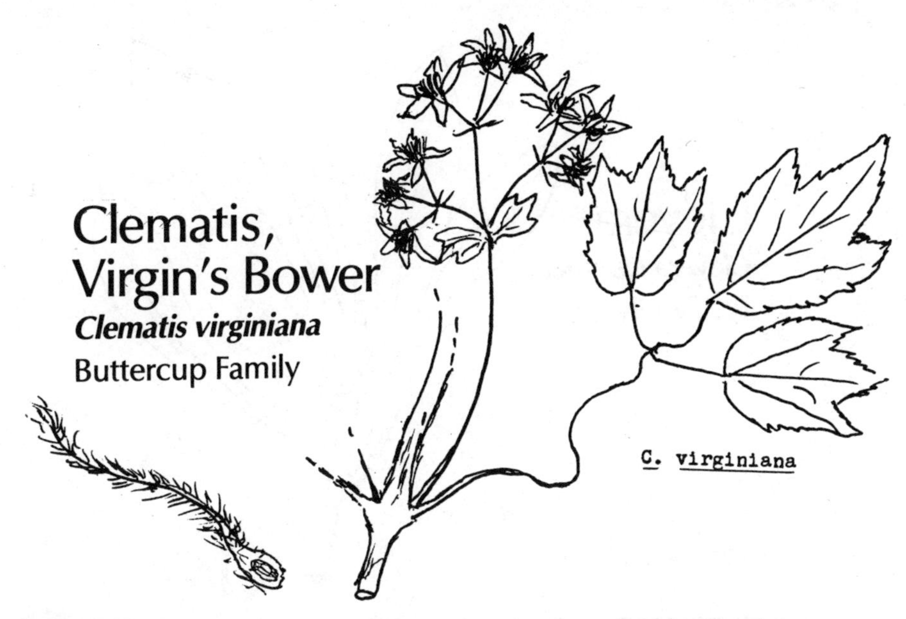

Clematis, Virgin's Bower

Clematis virginiana

Buttercup Family

GROWTH HABIT: Hardy, deciduous, woody vine, climbing by clasping petioles

LEAVES: Opposite, leaflets 3, each ovate, acuminate and toothed or lobed, 1-3″ long

FLOWERS: Numerous in axillary panicles, no petals, sepals white, petaloid and 1/2″ long, producing mass effect

FRUITS: Numerous achenes developing in center of flower, each with a long, persistant, plumose style

SEEDS: 1 per achene

LIGHT: Lots of sun

SOIL: Silt or clay loam

MOISTURE: Moist or wet

SEASONAL ASPECT: Flowering time in late summer

ZONE: 1, 2, 3

USES: Fence, low border, trellis

CARE: None

PROPAGATION BY: Seedlings

NATIVE OF: e. United States

OTHER: *C. paniculata (C. dioscoreifolia)*, flowers white, fragrant; leaflets not toothed; somewhat more hardy and more floriferous than above.

C. *armandii*, flowers white, leaflets much larger than above, 3-6″ long, and evergreen. Some barely climbing, deciduous hybrids with white to red to blue large, showy flowers have been developed.

Cleyera

Cleyera japonica

Tea Family

GROWTH HABIT: Evergreen shrub attaining large or small tree size if not pruned
LEAVES: Alternate, elliptic to obovate, entire, glossy, often with reddish mid-veins
FLOWERS: Small, white, slightly fragrant, axillary
FRUITS: Red berries 1/2" wide
SEEDS: Few per berry
LIGHT: Sun or partial shade
SOIL: Fertile loam
MOISTURE: Moist but well drained
SEASONAL ASPECT: When produced, the red berries in fall are very attractive
ZONE: 1, 2
USES: Group or unit planting, shrub border and hedge
CARE: Very little pruning required
PROPAGATION BY: Cuttings
NATIVE OF: Japan to India
OTHER: *Ternstroemia gymnanthera* is closely related, similar to and often confused with *Cleyera.* The leaves on *Ternstroemia* are clustered toward the twig tips and the flowers and fruits are yellow. *Eurya japonica* is similar but has unisexual flowers.

Climbing Hydrangea

Schizophragma integrifolia

Saxfrage Family

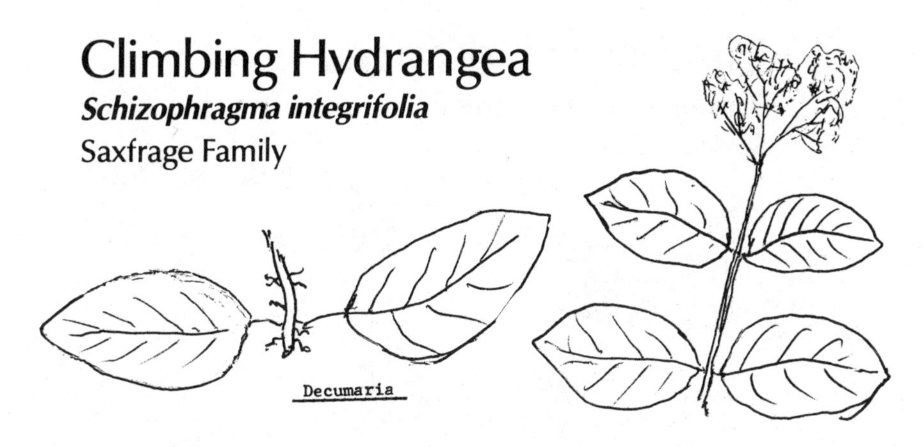

GROWTH HABIT: Woody, semievergreen vine, high-climbing by aerial roots
LEAVES: Opposite, 1-4″ long, entire, or toothed
FLOWERS: White, regular, numerous and borne in terminal clusters
FRUITS: Capsule, 1/3″ long, thin walled and 7-10 celled
SEEDS: Very small and numerous
LIGHT: Shade or a little sun
SOIL: Rich sandy or silty loam
MOISTURE: Wet to moist
SEASONAL ASPECT: None
ZONE: 1, 2, 3
USES: In low, damp or wet place around tree or tall rock
CARE: None
PROPAGATION BY: Cuttings or rooted stems
NATIVE OF: s.e. United States
OTHER: *Hydrangea anomala*; Japan; deciduous, big petioled cordate leaves; flat-topped clusters of small white flowers; climbing by aerial roots.

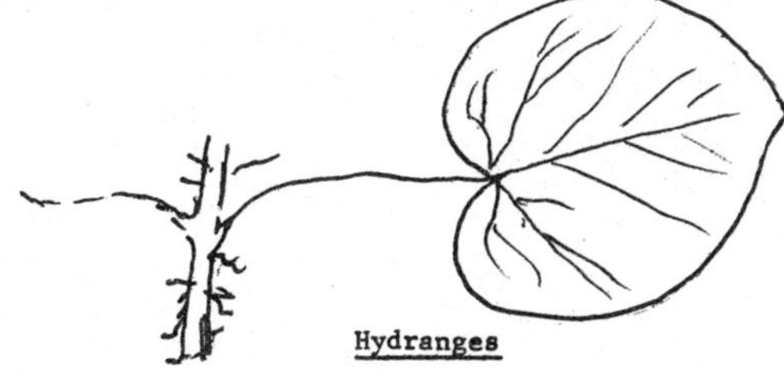

Climbing Hydrangea

Schizophragma integrifolia

Saxfrage Family

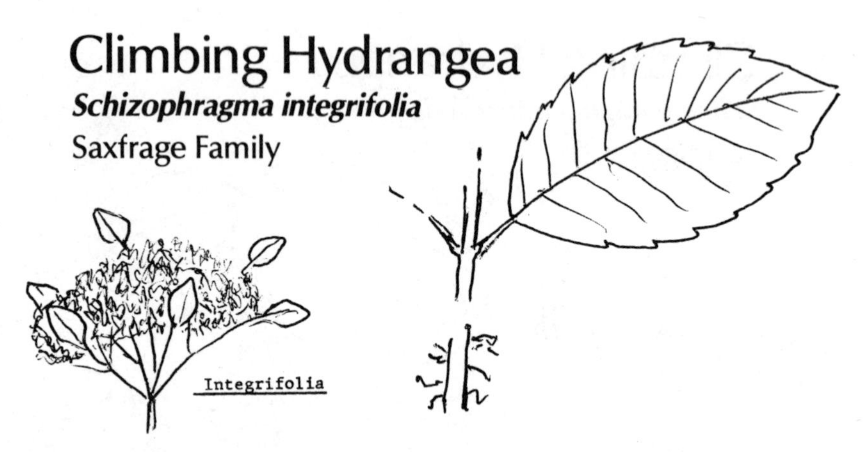

GROWTH HABIT: Deciduous vine climbing by aerial roots
LEAVES: Coarsely toothed and to 4″ long
FLOWERS: In rounded clusters to 10″ across, white
FRUITS: Small capsule
SEEDS: Tiny and many
LIGHT: Partial shade
SOIL: Rich loam
MOISTURE: Moist
SEASONAL ASPECT: Summer flowers
ZONE: 1, 2
USES: On trees
CARE: None
PROPAGATION BY: Seedlings
NATIVE OF: China
OTHER: *S. hydrangeoides*; Japan; similar to above but with somewhat smaller flower heads.

Star Jasmine, Confederate Jasmine

Trachelospermum jasminoides

Dogbane Family

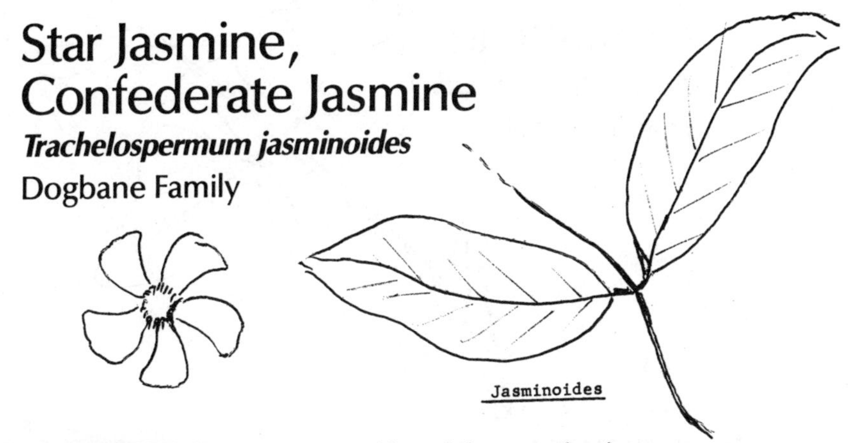

GROWTH HABIT: Evergreen, weakly twining woody vine
LEAVES: Opposite, elliptic, short pointed, glossy, 2-3″ long and with milky juice
FLOWERS: White, about 1″ across, very fragrant, 5-parted and in axillary and terminal clusters
FRUITS: Very thin paired pods to 6″ long
SEEDS: Many and with long silky hairs
LIGHT: Partial shade
SOIL: Most any fertile type
MOISTURE: Moist to dry
SEASONAL ASPECT: Flowers mid-spring to mid-summer
ZONE: 1, 2, 3
USES: Post, trellis, fence, walk
CARE: Training to support
PROPAGATION BY: Cuttings, layering
NATIVE OF: China
OTHER: *T. asiaticum*; Japan, Korea; evergreen; flowers cream or yellowish; possibly a little more hardy than the above.
T. difforme; e. United States; tardily deciduous; leaves varying on some plant from linear to ovate; flowers small, yellowish and not conspicuous.

Coral-Bean

Erythrina crista-galli

Legume Family

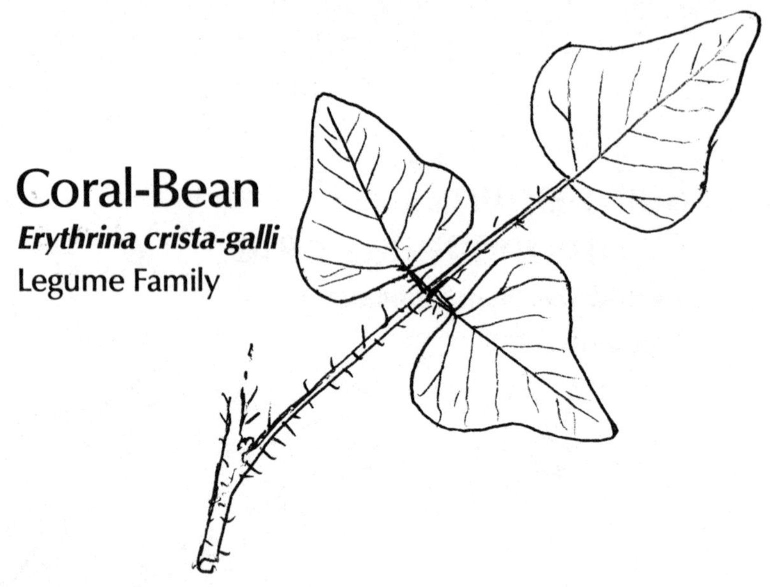

GROWTH HABIT: Deciduous, spiny-stemmed shrub
LEAVES: Trifoliate; leaflets entire and hairless; leaf-stalks spiny
FLOWERS: Bright scarlet, to 2″ long and in terminal spikes
FRUITS: Bean-type pod constricted between seeds
SEEDS: 2-5 and brightly colored
LIGHT: Lots of sun
SOIL: Sandy loam
MOISTURE: Dry
SEASONAL ASPECT: Early summer flowers
ZONE: 1, 2
USES: Use where flowers can be seen
CARE: Removal of dead branches following blooming
PROPAGATION BY: Seeds, seedlings
NATIVE OF: Brazil
OTHER: New branches arise from the roots following blooming

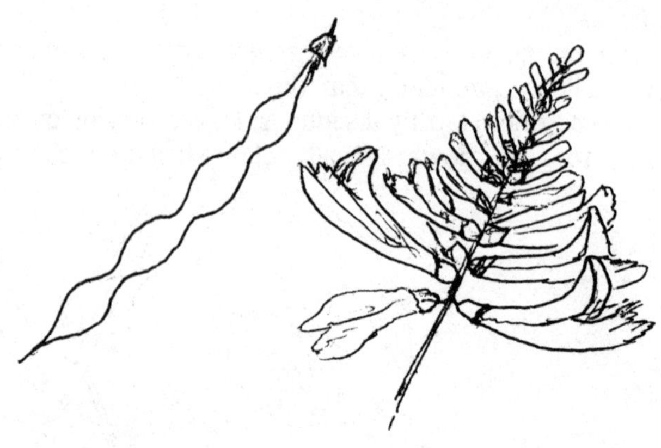

Coral Berry, Indian Currant

Symphoricarpos orbiculatus

Honeysuckle Family

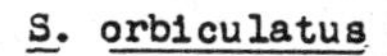

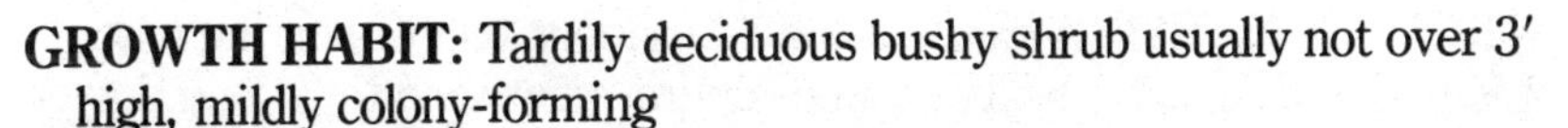

GROWTH HABIT: Tardily deciduous bushy shrub usually not over 3′ high, mildly colony-forming
LEAVES: Opposite, mostly ovate, to about 1″ long
FLOWERS: Very small, greenish, inconspicuous, mid-summer
FRUITS: Coral-like, purplish berries in small clusters
SEEDS: 2, one or both often abort
LIGHT: Partial shade
SOIL: Tolerant to most types
MOISTURE: Moist
SEASONAL ASPECT: Fall fruits
ZONE: 2, 3
USES: Rockery, bank, informal shrub border
CARE: None
PROPAGATION BY: Suckers
NATIVE OF: e. United States
OTHER: *S. X chenaultii* (*S. orbiculatus X S. microphyllus*); a low spreading type.

Coral-Vine, Corallita

Antigonon leptopus

Smartweed Family

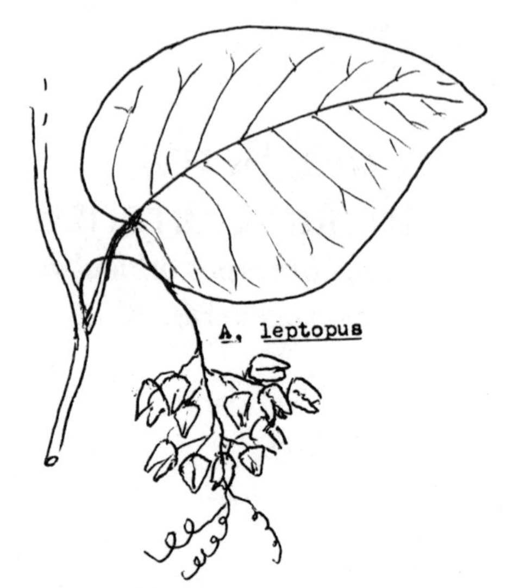

GROWTH HABIT: A barely woody deciduous tendril-bearing vine
LEAVES: Alternate, heart-shaped, entire, veiny and to 3″ long, lower larger
FLOWERS: Bright pink, about l" long, 5 parted and in tendril-tipped cluster
FRUITS: 3-angled achene, 1/4″ long
SEEDS: 1
LIGHT: Full sun
SOIL: Fertile
MOISTURE: Dry to moist
SEASONAL ASPECT: Summer flowers
ZONE: 1
USES: Wall, trellis, fence
CARE: None
PROPAGATION BY: Seeds or seedling
NATIVE OF: Mexico
OTHER: White Corallita, Christmas-Vine; *Porana paniculata*; Morning-Glory Family; India, is sometimes mistaken for the white flowered variety of *Antigonon*; *Porana* has no tendril at tip of flower clusters and a capsule of several seeds.

Cotoneaster

Cotoneaster spp.

Rose Family

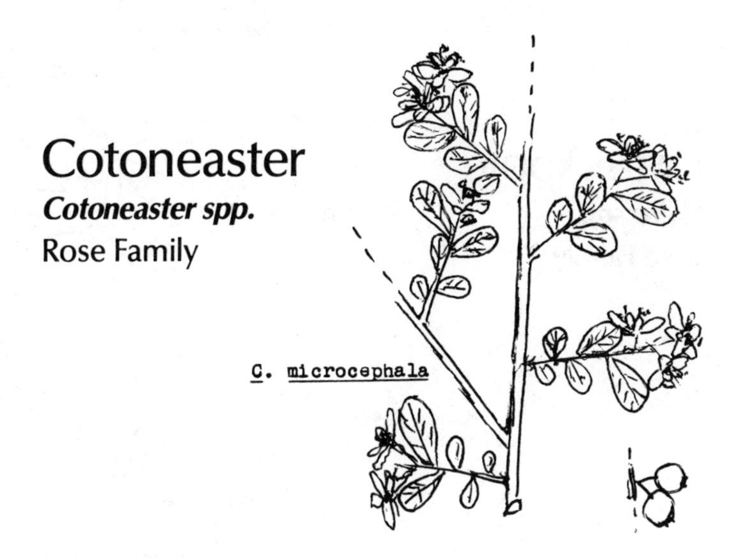

GROWTH HABIT: Evergreen and deciduous shrubs of different sizes and growth habits; 3-12′ high

LEAVES: Alternate, numerous, small, entire, 1/2 - 1″ long

FLOWERS: Small but numerous, white or pink, solitary or in clusters

FRUITS: Red or blackish, 1/4″ wide

SEEDS: Mostly 3

LIGHT: Full sun

SOIL: Fertile loam

MOISTURE: Moist

SEASONAL ASPECT: Spring flowers

ZONE: 2, 3

USES: Foundation, unit arrangement, hedge, border

CARE: Guard against fire blight, lace insects and red spiders

PROPAGATION BY: Cuttings

NATIVE OF: China, Himalayas, Siberia

OTHER: Evergreen:

C. *microcephala*; Himalayas; to 3′ high, leaves about 1/4″ long; flowers white and 1-3 together.

C. *rotundifolia*; Himalayas; to 10′ high; leaves about 1/2″ long; flowers white and solitary.

Deciduous:

C. *adpressum*; w. China; prostrate shrub; secondary branches borne irregularly on main branches.

C. *horizontalis*; China; prostrate shrub; secondary branches borne either on the right or left side of main branches.

C. *dielsiana*; China; erect shrub; flowers pink.

C. *racemiflora*; Europe and N. Africa; erect shrub; flowers white.

Crab-Apple

Malus spp.

Rose Family

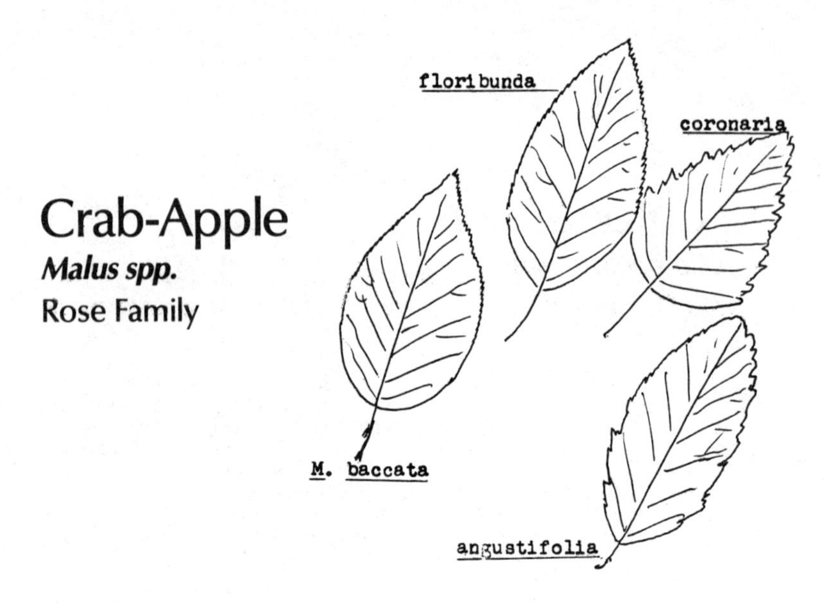

GROWTH HABIT: Deciduous small trees, or sometimes shrubs, to 25′ high, grown mostly for blossoms
LEAVES: Alternate, ovate to oblong to elliptic, entire, toothed or shallowly lobed, green or purplish-red
FLOWERS: Abundantly produced, 3/4 - 1 1/2″ wide, white, pink or rose, single or semidouble
FRUITS: Apples about 1″ wide, green, yellow or red; edible in jelly
SEEDS: Several
LIGHT: Full or partial sun
SOIL: Fertile
MOISTURE: Moist
SEASONAL ASPECT: Flowering just before leaves appear
ZONE: 1, 2, 3
USES: Decoration, sidewalk, street, freestanding
CARE: None
PROPAGATION BY: Seedlings and cuttings
NATIVE OF: China, Japan, Siberia and e. America
OTHER: The below and other species are cultivated in several varieties.
M. baccata; Siberian Crab; Siberia and N. China; flowers white.
M. floribunda; Showy Crab; China; flowers rose- pink, fading white.
M. coronaria; Wild Sweet Crab; e. United States; leaf shallowly toothed; flowers rose colored.
M. angustifolia; Southern Crab; e. United States; leaves toothed and lobed; flowers fragrant and rose colored.

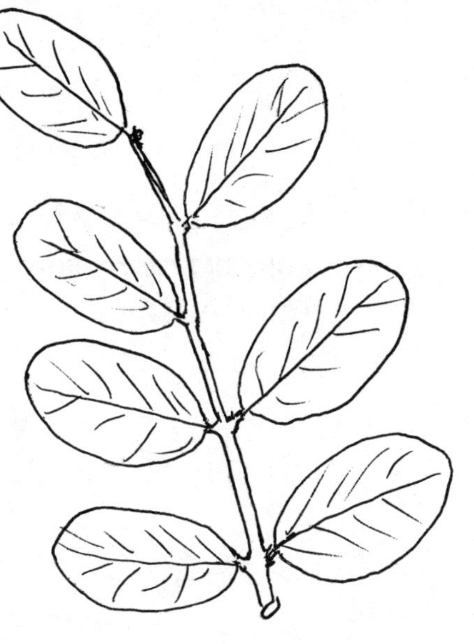

Crepe Myrtle

Lagerstroemia indica

Loosestrife Family

GROWTH HABIT: Deciduous tree or shrub; trunk irregular in diameter with smooth greenish bark that exfoliates leaving a mottled pattern
LEAVES: 1-2″ long, elliptic and pubescent on the veins beneath
FLOWERS: White, pink, purple or red, crinkled and to 1 1/2″ wide
FRUITS: Woody capsule
SEEDS: Winged
LIGHT: Full sun
SOIL: Tolerant
MOISTURE: Moist
SEASONAL ASPECT: Summer flowers, autumn color. Leaves of white flowered varieties show only yellow in fall.
ZONE: 1, 2, 3
USES: Street, sidewalk, freestanding
CARE: None
PROPAGATION BY: Cuttings
NATIVE OF: Asia and n. Australia
OTHER: Many varieties in size and color.

Cross-Vine

Bignonia capreolata

Trumpet Creeper Family

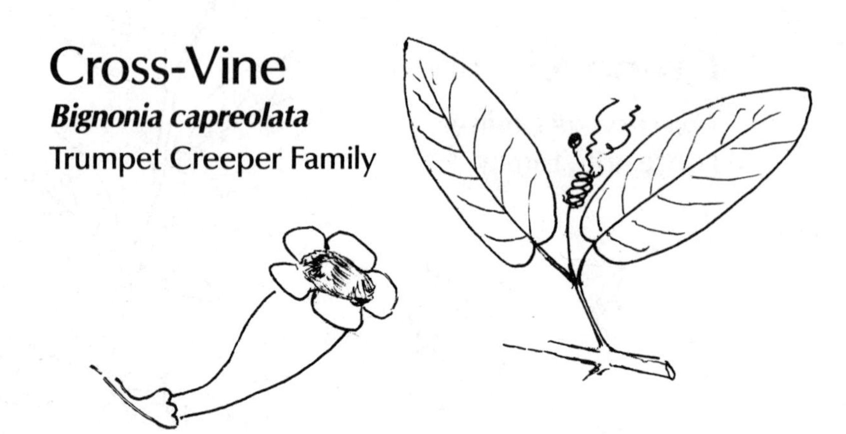

GROWTH HABIT: Woody, nearly evergreen vine climbing by means of tendrils

LEAVES: Opposite, compound with 2 leaflets and a tendril, leaflets oblong to elliptic, entire and heart-shaped at base; tendril branches with terminal adhesive disks

FLOWERS: 2″ long, 5-lobed with open center, yellow and red

FRUITS: Flat capsule to 6″ long

SEEDS: Many and broadly winged

LIGHT: Full or partial sun

SOIL: Fertile heavy soil

MOISTURE: Moist

SEASONAL ASPECT: Late spring flowers

ZONE: 1, 2 (3)

USES: Arbor, trellis, fence, tree, screen

CARE: None

PROPAGATION BY: Seedlings

NATIVE OF: e. United States

OTHER: The name *Anisostichus* is now used, but many will yet know it only by the old name *Bignonia.* The name Cross-Vine comes from a pattern in the wood as seen in a stem cross section. A Humming Bird attractant.

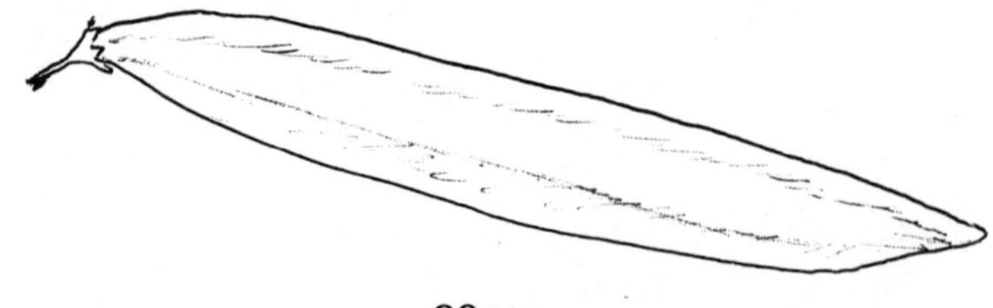

Cryptomeria

Cryptomeria japonica

Taxodium Family

GROWTH HABIT: Evergreen tree or shrub with dense foliage and whorled branches
LEAVES: Short, narrow, sharp pointed and keeled
FLOWERS: Male as yellowish catkins in spring; female as 1″ long cones
FRUITS: Somewhat leathery, reddish-brown cones in autumn
SEEDS: Narrowly winged nutlets, several per cone
LIGHT: Full or partial sun
SOIL: Quite tolerant
MOISTURE: Moist to dry
SEASONAL ASPECT: None
ZONE: 1, 2, 3
USES: Dependent on size of the variety used
CARE: None
PROPAGATION BY: Seedlings
NATIVE OF: China

Cupseed

Calycocarpum lyoni

Moonseed Family

GROWTH HABIT: Deciduous high-climing woody vine
LEAVES: Large, to 1′ long and 3-7 lobed
FLOWERS: Numerous, very small, without petals and in panicles
FRUITS: Black drupe to 1″ long
SEEDS: 1
LIGHT: Partial shade
SOIL: Fertile loam or alluvium
MOISTURE: Moist to wettish
SEASONAL ASPECT: None
ZONE: 1, 2
USES: Arbor, trellis, tree
CARE: None
PROPAGATION BY: Seeds
NATIVE OF: Ind. to Okla. and south to La. and Fla.

Clove Currant

Ribes odoratum

Saxifrage Family

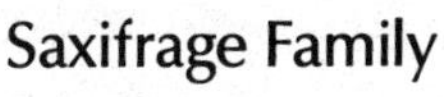

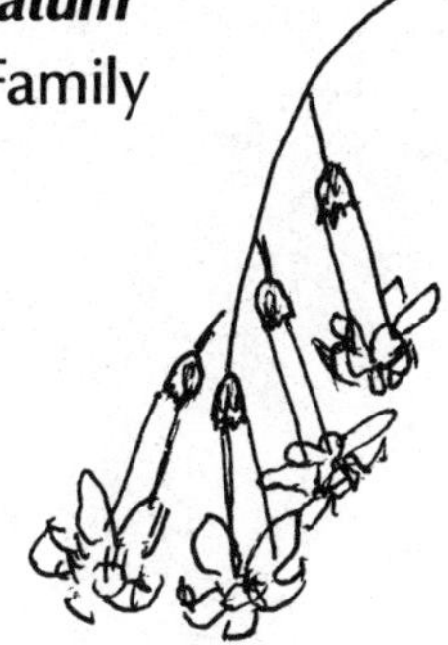

GROWTH HABIT: Deciduous shrub to 5′
LEAVES: Ovate or broader, to 2″ wide and 3-5 lobed
FLOWERS: Yellow, fragrant, 1/2″ long and in racemes
FRUITS: Black berry
SEEDS: Many
LIGHT: Lots of sun
SOIL: Good
MOISTURE: Moist
SEASONAL ASPECT: Spring flowers
ZONE: 2, 3
USES: Border
CARE: None
PROPAGATION BY: Seeds
NATIVE OF: central United States
OTHER: Varieties have been developed, including the Crandall Currant.

Cycad

Cycas revoluta

Cycad Family

GROWTH HABIT: Palmlike evergreen shrub with short, stout trunk, to 6′ high

LEAVES: To 3″ long, one-pinnate with many narrow divisions, lusterous

FLOWERS: Densely hairy structures atop stem

FRUITS: Smooth, orange colored, egg shaped structures about 1 1/2″ long

SEEDS: Few

LIGHT: Shade or some sun

SOIL: Sandy loam

MOISTURE: Moist or fairly dry

SEASONAL ASPECT: None

ZONE: 1, and protected locations in 2

USES: Lends tropical appearance to terrace, patio or garden

CARE: Protect from frost and salt spray

PROPAGATION BY: Seeds, seedlings and stem offsets

NATIVE OF: Africa

Cypress

Cupressus macnabiana

Cypress Family

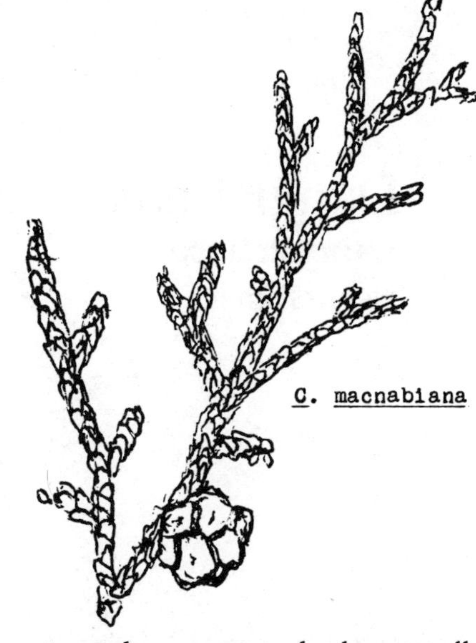

GROWTH HABIT: Broad or cone-topped, evergreen shrub or small tree with branch tips forming flat sprays.
LEAVES: Scale-like and appressed, needle-like on juveniles
FLOWERS: Minute and terminal; male and female on separate branches
FRUITS: 1″ wide globose cones with 6-8 thick-ended scales
SEEDS: Winged
LIGHT: Full or partial sun
SOIL: Fertile loam
MOISTURE: Dry or moist
SEASONAL ASPECT: None
ZONE: 1, 2, 3
USES: Many, depending on variety
CARE: Little or none
PROPAGATION BY: Seeds and seedlings
NATIVE OF: U.S. west coast
OTHER: One of the shrub varieties has twigs that are golden-tipped Arizona Cypress, *C. arizonica*, resembles Red Cedar and is sometimes planted for Christmas trees.

Dangleberry

Gaylussacia spp.

Heath Family

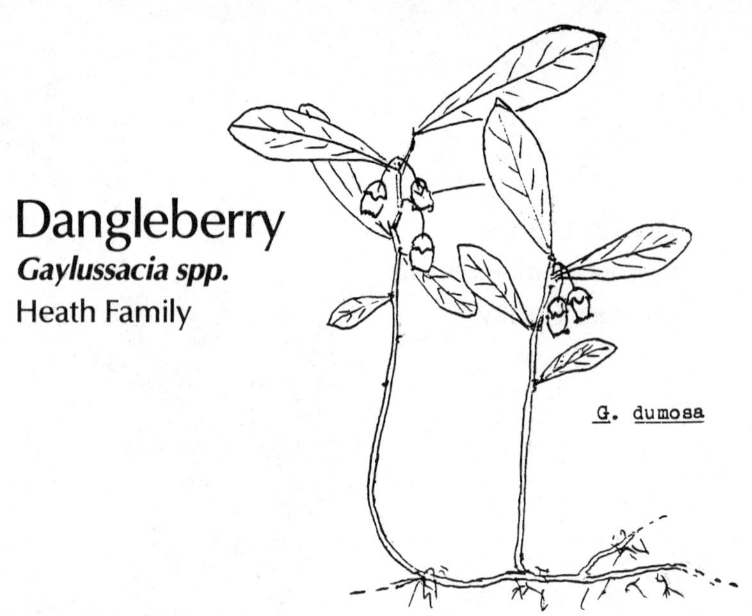

GROWTH HABIT: Low, deciduous shrubs to 2′ high, freely branched and colony forming

LEAVES: Alternate, elliptic to oblanceolate, to 1 1/2″ long and minutely resinous dotted beneath

FLOWERS: White or tinged with pink, urn-shaped and borne in small clusters

FRUITS: 10-celled, 10-seeded black berries

SEEDS: 10

LIGHT: Partial shade

SOIL: Sandy or clay loam

MOISTURE: Moist to dry

SEASONAL ASPECT: None

ZONE: 1, 2, 3

USES: Open ground cover, interest

CARE: None

PROPAGATION BY: Colony division

NATIVE OF: e. United States

OTHER: Box Huckleberry; *G. brachycera*; leaves finely toothed; fruit glandular pubescent, black.

Dwarf Huckleberry; *G. dumosa*; leaves entire; fruit black.

Dangleberry, *G. frondosa*; leaves entire; plant somewhat larger than the above; fruit bluish.

Daphne

Daphne mezereum

Mezereum Family

GROWTH HABIT: Tardily deciduous, almost succulent shrub to 3 1/2′ high
LEAVES: Alternate, oblong or oblanceolate, 2-3″ long and glabrous
FLOWERS: Lilac purple, fragrant, in clusters along stem before the leaves
FRUITS: Scarlet and 1/4″ long
SEEDS: 1
LIGHT: Partial shade
SOIL: Tolerant to most well-drained, neutral types
MOISTURE: Moist
SEASONAL ASPECT: Early spring flowers and fragrance
ZONE: 1, 2
USES: Interest, specimen, unit arrangement, occasional flowering shrub
CARE: Keep the roots cool
PROPAGATION BY: Cuttings and seedlings
NATIVE OF: Europe and w. Asia
OTHER: A number of species and varieties are available, most of which conform generally to the description above. A few are evergreen. One problem after plants are established sometimes is sudden and unsatisfactorily explained death. Berries toxic.

Daubentonia

Daubentonia punicea

Legume Family

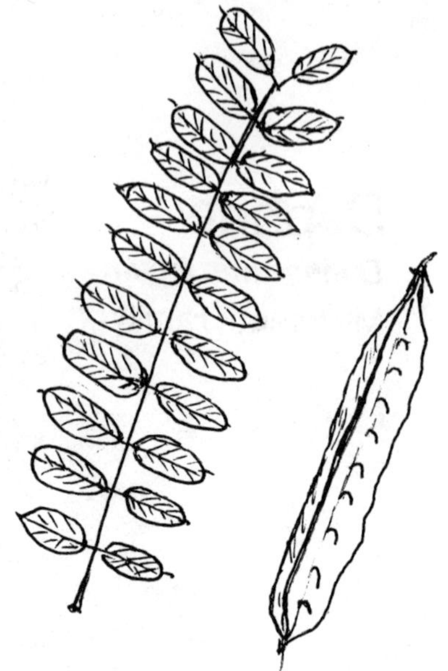

GROWTH HABIT: Deciduous shrub to 8; high
LEAVES: Alternate, pinnate, 12-40 leaflets, each entire and 1/2 - 1″ long
FLOWERS: 3/4″ long, deep orange to reddish-purple in crowded racemes to 4″ long
FRUITS: 3-4″ long plump, brownish, 4-winged pods
SEEDS: Several separated from one another by partitions
LIGHT: Full sun
SOIL: Tolerant
MOISTURE: Moist
SEASONAL ASPECT: Summer flowers
ZONE: 1, 2
USES: Occasional flowering shrub, interest
CARE: None
PROPAGATION BY: Seeds
NATIVE OF: S. America, naturalized in s.e. United States

Deerberry, Gooseberry, Squaw-Huckleberry

Polycodium stamineum
(*Vaccinium* by some authors)
Heath Family

GROWTH HABIT: Much branched, deciduous shrub to 3′ high
LEAVES: Alternate, elliptic to oblanceolate, usually whitened, especially beneath, entire and to 3″ long, much smaller back from twig tip
FLOWERS: White, corolla lobes as long or longer than tube
FRUITS: Berry green, purplish, pinkish or yellowish, mostly with whitened surface; generally insipid but an occasional plant with good fruit
SEEDS: Many
LIGHT: Three-fourths sun
SOIL: Sandy
MOISTURE: Dry to moist
SEASONAL ASPECT: Late spring flowers or late summer fruit
ZONE: 1, 2, 3
USES: Interest
CARE: None
PROPAGATION BY: Seedlings
NATIVE OF: e. United States
OTHER: Selection might well render this species of economic value for its fruit.

Deodar Cedar

Cedrus deodara

Pine Family

GROWTH HABIT: Large evergreen tree
LEAVES: Alternate, but on older branches become closely clustered on short spur twigs; needle-shaped and about 1 1/2″ long
FLOWERS: Male in yellowish catkins; female in cones
FRUITS: Woody cones about 4″ long that at maturity shed their scales except for the terminal dozen or so that fall as a unit
SEEDS: With large thin wing
LIGHT: Full sun
SOIL: Fertile loam
MOISTURE: Moist
SEASONAL ASPECT: None
ZONE: 1, 2, 3
USES: Freestanding, framing, background
CARE: None
PROPAGATION BY: Seedlings
NATIVE OF: Himalayas and Asia Minor
OTHER: *C. atlantica,* leaves shorter, twigs slender and drooping. The several cone-tip scales falling as a unit somewhat resemble a rose and are used in dry arrangements.

Deutzia

Deutzia scabra

Saxifrage Family

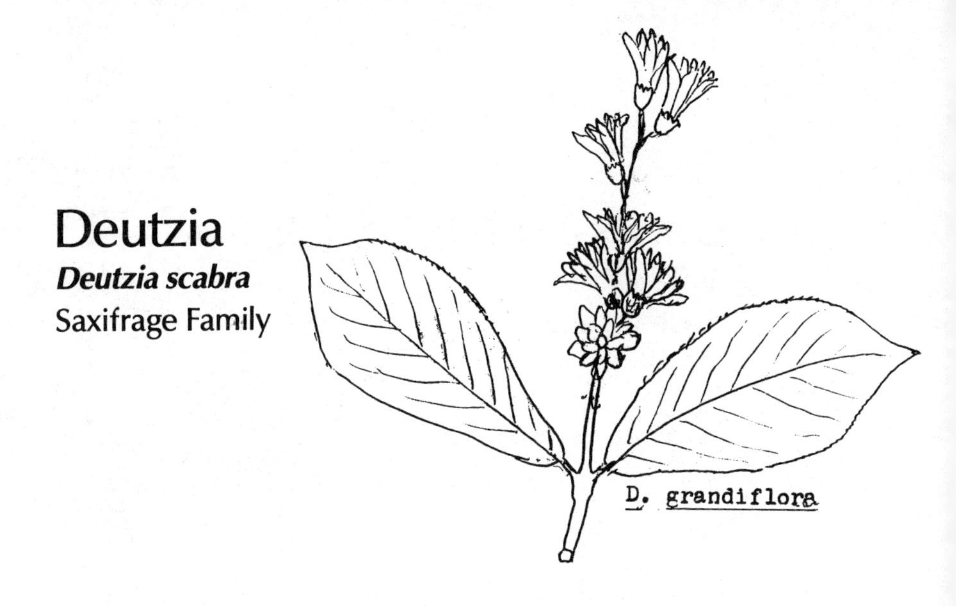

GROWTH HABIT: Hardy, deciduous shrub with numerous erect branches and brownish, peeling bark; to 6′ high

LEAVES: Opposite, ovate, entire to 3″ long and rough with stellate hairs

FLOWERS: White to pinkish, 1/3″ long and produced abundantly in short axillary panicles

FRUITS: 3-5-celled capsule

SEEDS: Numerous and tiny

LIGHT: Full or partial sun

SOIL: Rich, loam type

MOISTURE: Moist

SEASONAL ASPECT: Flowering time in late spring

ZONE: 1, 2, 3

USES: Informal shrub border, occasional shrub, specimen

CARE: Little or no pruning

PROPAGATION BY: Cuttings

NATIVE OF: China and Japan

OTHER: The above exists in several varieties.

D. grandiflora; China; to 6′ high, earliest to bloom and with largest flowers.

D. X kalmiaeflora (*D. purpurascens X D. parviflora*); to 5′ high, flowers white inside, carmine outside, almost 1″ wide.

Devil's-Walking-Stick

Aralia spinosa

Ginseng Family

GROWTH HABIT: Little branched, deciduous shrub or small tree, spiny and to 25′ high
LEAVES: Great in size, twice pinnate, 2′ or more long; leaflets ovate, 2-3″ long and finely toothed
FLOWERS: Small and collected in great terminal clusters to 3′ wide
FRUITS: Small purplish berry
SEEDS: 1-few
LIGHT: Partial shade
SOIL: Fertile loam
MOISTURE: Moist
SEASONAL ASPECT: Whitish flower mass in midsummer or the purple fruits in early fall
ZONE: 1, 2, 3
USES: Oddity, interest
CARE: None
PROPAGATION BY: Seedlings or root sprouts
NATIVE OF: s.e. United States

Dichondra

Dichondra carolinensis

Morning-Glory Family

GROWTH HABIT: Small prostrate evergreen herb
LEAVES: Small, almost circular and to 1″ wide
FLOWERS: Greenish-white, 5-parted and 1/2″ wide
FRUITS: Small capsule
SEEDS: Several
LIGHT: Partial shade
SOIL: Most any
MOISTURE: Moist
SEASONAL ASPECT: None
ZONE: 1, 2
USES: Lawn, ground cover
CARE: None
PROPAGATION BY: Seeds, sod
NATIVE OF: s.e. United States

Maryland Dittany, Stone-Mint

Cunila origanoides

Mint Family

GROWTH HABIT: Deciduous subshrub to 1 high
LEAVES: Opposite, ovate, 1″ long, toothed, fragrant
FLOWERS: Pinkish, 1-lipped, 1/4″ long and in terminal clusters
FRUITS: 4 nutlets per flower
SEEDS: 1 per nutlet
LIGHT: Shade, beneath high canopy
SOIL: Clay loam rich in humus
MOISTURE: Moist
SEASONAL ASPECT: Early autumn flowers
ZONE: 2
USES: Borders
CARE: None
PROPAGATION BY: Seedlings
NATIVE OF: e. United States

Dog-Laurel

Leucothoe axillaris

Heath Family

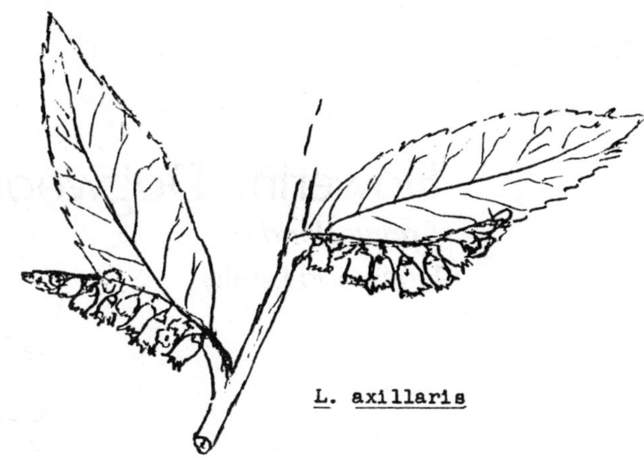

GROWTH HABIT: Evergreen shrub to 4′ with arching branches

LEAVES: Appearing 2-ranked, lanceolate to narrowly elliptic, finely toothed, glossy green above and to 5″ long

FLOWERS: White or tinted, tubular but constricted at the tip, 1/4″ long and in dense axillary racemes

FRUITS: Depressed capsules

SEEDS: Flat

LIGHT: Shade

SOIL: Sandy or clay loam

MOISTURE: Moist to wet

SEASONAL ASPECT: Spring flowers

ZONE: 1, 2, 3

USES: Low border, blender

CARE: Slowly forms colonies

PROPAGATION BY: Roots or cuttings

NATIVE OF: e. United States

OTHER: *L. editorum* (*L. catesbaei, L. fontanesiana*) has drooping racemes of fragrant white flowers. Leaves turn purplish in winter; zone 3.

Flowering Dogwood

Cornus florida

Dogwood Family

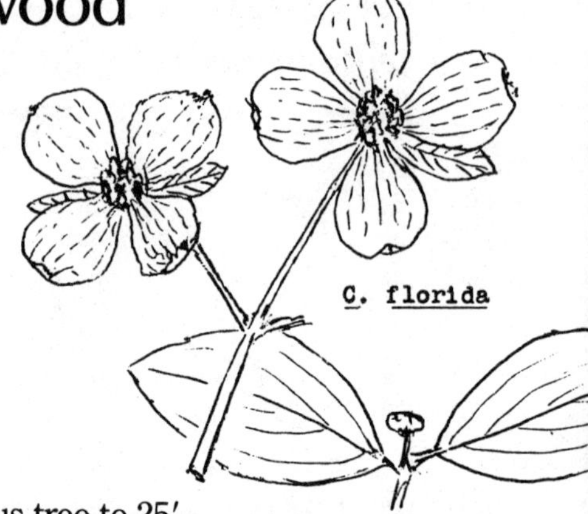

GROWTH HABIT: Small deciduous tree to 25′
LEAVES: Opposite, ovate, 2-5″ long
FLOWERS: Small, greenish-yellow, sessile and crowded; surrounded by four white bracts, bracts and flower cluster 2 1/2 - 3 1/2″ wide
FRUITS: Red, berry-like and 1/2″ long
SEEDS: 2
LIGHT: Partial shade as beneath high canopy
SOIL: Fertile loam
MOISTURE: Moist
SEASONAL ASPECT: Flowering as the leaves appear
ZONE: 1, 2, 3
USES: Freestanding, street decoration, specimen
CARE: Little if any
PROPAGATION BY: Seedlings
NATIVE OF: e. United States
OTHER: The wood is very hard and was extensively used by the textile industry for shuttles. *C. nuttallii*; a west coast species with somewhat larger flowers.

Cornus, Red Osier

Cornus spp.

Dogwood Family

GROWTH HABIT: Deciduous shrubs, mostly with red twigs
LEAVES: Opposite, ovate or broadly so, entire, veins curving toward tip
FLOWERS: Small, whitish and in loose terminal clusters and without showy involucral bracts
FRUITS: Blue, white or pinkish 1/4″ wide berries
SEEDS: 2
LIGHT: Partial shade
SOIL: Rich, alluvial or loam
MOISTURE: Wet, moist or dry
SEASONAL ASPECT: Red or purplish twigs after leaf-fall
ZONE: 1, 2, 3
USES: Occasional shrub
CARE: Some pruning
PROPAGATION BY: Seedlings
NATIVE OF: e. United States
OTHER: C. *amomum*; e. United States; young branches green to red, pith brown in two year old stems; fruit blue.

C. *racemosa*; e. United States; young branches red, becoming gray, fruit white.

Swamp Dogwood; C. *stricta*; e. United States; young branches red to gray; pith white in two year old stems; fruit bluish.

Red Osier; C. *stolonifera*; n.e. United States; young branches red; pith white; fruit white; thicket-former if allowed. Other species, mostly introduced, are in the trade.

Dove Tree, Handkerchief Tree

Davidia involucrata

Nyssa Family

GROWTH HABIT: Deciduous tree
LEAVES: Alternate, toothed, cordate, to 6″ long and silky pubescent beneath
FLOWERS: Unisexual, without sepals or petals, very small and crowded; 1 female, many males per cluster subtended by large white bract
FRUITS: Pear-shaped, greenish and 1 1/2″ long
SEEDS: 1
LIGHT: Full sun
SOIL: Fertile
MOISTURE: Moist
SEASONAL ASPECT: Late spring flowers
ZONE: 1, 2, 3
USES: Specimen
CARE: Little if any
PROPAGATION BY: Seedlings and cuttings
NATIVE OF: China
OTHER: This plant is spectacular when in bloom. The white bracts subtending the flowers are up to 6″ wide and suggest handkerchiefs hanging out to dry, hence the sometimes used name Handkerchief-Tree. Does not flower every year.

Dutchman's Pipe, Pipe-Vine

Aristolochia macrophylla

Birthwort Family

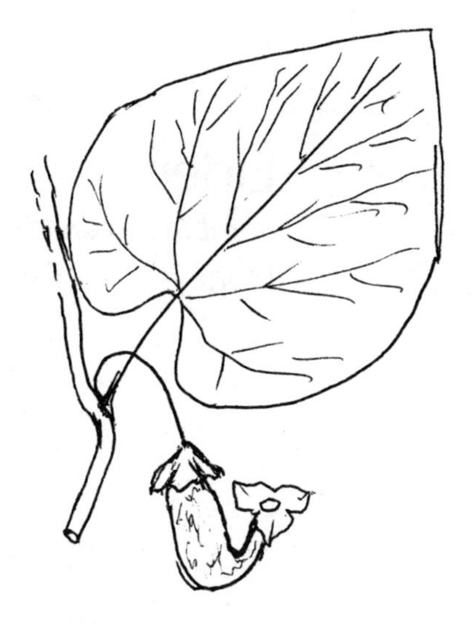

GROWTH HABIT: Woody twining vine; deciduous
LEAVES: Alternate, or crowded on spur twigs, large, to 10″ long, broadly heart-shaped and soft downy
FLOWERS: Green or maroon with long curved calyx tube
FRUITS: Pipe-shaped, to 3″ long, a capsule
SEEDS: Flattened and many
LIGHT: Half shade or more
SOIL: Fertile
MOISTURE: Moist to wet
SEASONAL ASPECT: None
ZONE: 2
USES: Interest, oddity, trellis, porch vine
CARE: Provide support
PROPAGATION BY: Seeds
NATIVE OF: central United States

Elliottii

Elliottii racemosa

Heath Family

GROWTH HABIT: Rare, deciduous shrub to 8′
LEAVES: Alternate, elliptic, shallowly toothed and to 5″ long
FLOWERS: Small, white and in long terminal racemes
FRUITS: Capsule
SEEDS: Several
LIGHT: Three-fourths sun
SOIL: Sandy, acid
MOISTURE: Moist
SEASONAL ASPECT: Summer flowers
ZONE: 2, 3
USES: Specimen
CARE: —
PROPAGATION BY: Cuttings, vernalized seeds
NATIVE OF: South Carolina and Georgia

American Elm and others

Ulmus americana

Elm Family

GROWTH HABIT: Large, spreading deciduous tree
LEAVES: Alternate, rough, ovate-oblong, sharply toothed, abruptly pointed, unequal sided at base and 3-6″ long
FLOWERS: Small, brownish, clustered and before the leaves
FRUITS: About 1/2″ long and winged all around
SEEDS: 1 per fruit
LIGHT: Full or partial sun
SOIL: Fertile loam
MOISTURE: Stream banks to dry roadsides
SEASONAL ASPECT: None
ZONE: 1, 2, 3
USES: Street, home, garden and park
CARE: None
PROPAGATION BY: Seedlings and cuttings
NATIVE OF: e. United States
OTHER: Once widely planted but now greatly restricted because of Dutch Elm disease.

Winged Elm, *U. alata*, e. United States; leaves 1/2 - 3″ long, twigs often developing two opposing corky wings with each becoming wider than the twig; fruits smaller and more narrowly winged than American Elm.

Chinese Elm, *U. parvifolia*; leaves also small as above but singly-toothed; drops leaves early.

Slippery Elm, *U. rubra*, e. United States; leaves large, 4-8″ long and very rough above, fruit and wing 3/4″ wide. The inner bark was once used or chewed by baseball pitchers and the resulting mucilaginous saliva used on "spit balls".

Smooth-leaved Elm, *U. carpinifolia*, Europe, n. Africa; leaves 2-3″ long and smooth above; many varieties.

Dutch Elm, *U. hollandica*; similar to above but with leaves 3-5″ long; many varieties; origin unknown.

English Ivy

Hedera helix

Ginseng Family

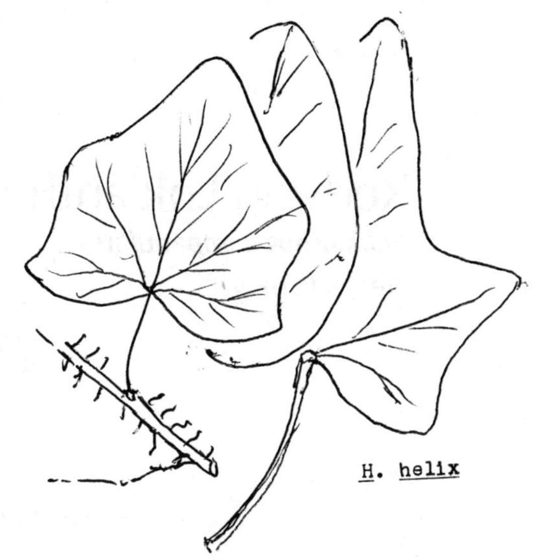

GROWTH HABIT: Extremely hardy and versatile evergreen vine climbing by numerous aerial roots or holdfasts to great heights

LEAVES: To 4″ long, mostly 3-5 lobed, margins entire, color somewhat mottled

FLOWERS: Small, greenish

FRUITS: Black, 1/4″ wide and in stalked clusters

SEEDS: Few

LIGHT: Sun or shade

SOIL: Tolerant

MOISTURE: Moist or dry

SEASONAL ASPECT: None

ZONE: 1, 2, 3

USES: Ground cover, over fence to form hedge, wall, tree, pot plants. Interesting if allowed to climb tree trunk for only about 4′

CARE: Containment

PROPAGATION BY: Cuttings

NATIVE OF: Europe

OTHER: Many varieties in the trade and nomenclature sometimes confusing. It has escaped in some areas. *H. canariensis*; with slender, usually reddish twigs and few, if any, aerial holdfasts. This is also cultivated in several varieties, especially variegated. Fruits and possibly other parts of plant toxic.

Redvein Enkianthus

Enkianthus campanulatus

Heath Family

GROWTH HABIT: Deciduous shrub to 6′ or more high
LEAVES: Elliptic, to 3″ long, finely toothed
FLOWERS: Yellow with darker veins, bell-shaped and to 1/2″ long
FRUITS: Capsule
SEEDS: Very small
LIGHT: Partial Shade
SOIL: Any acid type
MOISTURE: Moist
SEASONAL ASPECT: Early spring flowers (before the leaves) and scarlet leaves in fall
ZONE: 1, 2, 3
USES: Occasional shrub
CARE: None
PROPAGATION BY: Seedlings, cuttings
NATIVE OF: Japan
OTHER: *E. perulatus*; white flowered; to 6′ high; striking autumn foliage.

Erysimum

Erysimum linifolium

Mustard Family

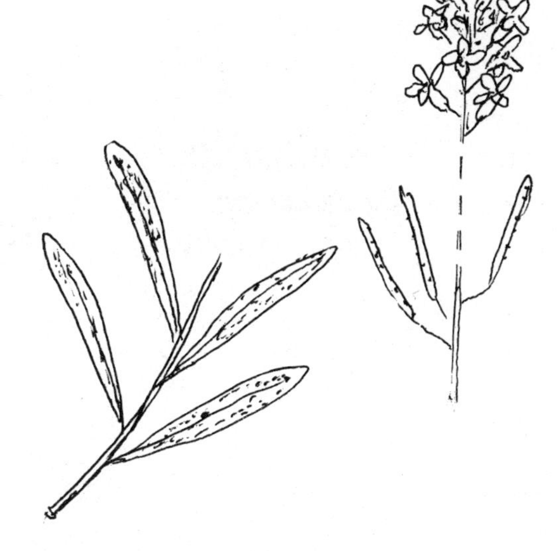

GROWTH HABIT: Woody based shrubby perennial, grayish, much-branched and to 2′ high
LEAVES: Alternate, mostly entire, narrow, to 2 1/2″ long and silvery pubescent
FLOWERS: Small, purple or paler, but showy on densely packed racemes raised 6-12″ above crown
FRUITS: 2″ long, 4-sided pod
SEEDS: Several
LIGHT: Full sun
SOIL: Fertile
MOISTURE: Moist but well drained
SEASONAL ASPECT: Late spring and early summer flowers
ZONE: 1, 2
USES: Border, patio, terrace
CARE: None
PROPAGATION BY: Seeds
NATIVE OF: Spain

Escallonia

Escallonia spp.

Saxifrage Family

GROWTH HABIT: Evergreen and tardily deciduous shrubs to 10′ high

LEAVES: Alternate or clustered evergreen, entire or glandular serrate, elliptic or broader, rather thick and to 3″ long

FLOWERS: White, pink or red, 5-parted, mostly in terminal arrangements

FRUITS: Top-shaped capsule

SEEDS: Many

LIGHT: Full or lots of sun

SOIL: Good porous loam

MOISTURE: Moist

SEASONAL ASPECT: Late spring flowers

ZONE: 1, 2, (3)

USES: Hedge, shrub border, unit arrangement

CARE: Pruning

PROPAGATION BY: Cuttings

NATIVE OF: s. America

OTHER: *E. X langleyensis* (*E. virgata X E. punctata*); evergreen shrub to 6′ high; leaves to 1″ long and very finely toothed; flowers deep pink and about 1/2″ wide.

E. punctata; Chile; evergreen shrub to 6′ high; leaves to 2″ long, broadest at or beyond the middle and sharply toothed; flowers almost 1″ wide, dark red and 2-3 together.

E. virgata; Chile; deciduous shrub to about 6′ high, bushy and cold hardy; leaves obovate, smooth and about 1/2″ long; flower white and 1/2″ wide.

Eucommia

Eucommia ulmoides

Eucommia Family

GROWTH HABIT: Medium-size deciduous tree
LEAVES: Alternate, toothed, pointed, rather like Elm
FLOWERS: Not showy, solitary, male and female on different trees, no sepals or petals
FRUITS: Winged nutlets
SEEDS: 1
LIGHT: Full or lots of sun
SOIL: Most any moderately fertile type
MOISTURE: Dry
SEASONAL ASPECT: None
ZONE: 1, 2, 3
USES: Specimen, freestanding, street
CARE: None
PROPAGATION BY: Seeds, summer cuttings
NATIVE OF: China

Eulalia

Miscanthus sinensis

Grass Family

GROWTH HABIT: Stout clump forming perennial, dying back in winter, 4-8′ high

LEAVES: To 3′ long, glabrous and narrow

FLOWERS: Terminal open panicles raised to the top or above the leaf clump

FRUITS: In 1-flowered spikelets

SEEDS: Very thin

LIGHT: Full sun for best growth

SOIL: Tolerant

MOISTURE: Moist

SEASONAL ASPECT: Summer panicles

ZONE: 1, 2, 3

USES: Interest, specimen

CARE: Removal of dead tops in winter

PROPAGATION BY: Rhizomes

NATIVE OF: China and Japan

OTHER: Forms with leaves striped with either white or yellow, or boldly barred with white are available. Wind-borne seeds sometimes account for new plants.

Euonymus

Euonymus spp.

Staff-Tree Family

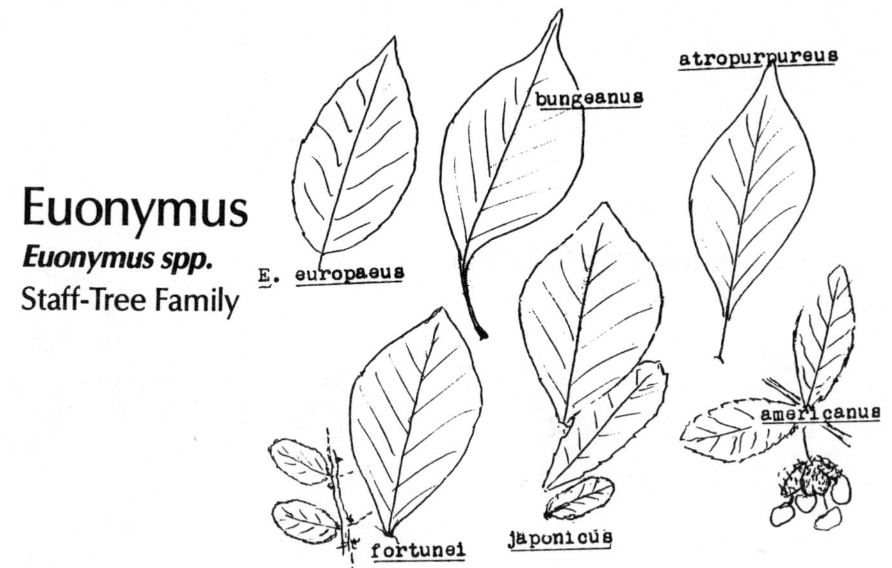

GROWTH HABIT: Those most common in the trade are evergreen shrubs or vines grown for their attractive foliage and fruits.

LEAVES: Opposite, usually with petiole and entire or toothed

FLOWERS: Small, greenish or purplish and in stalked axillary clusters

FRUITS: 3-5 lobed capsule 1/2″ or more wide, white, pink or red

SEEDS: Dark, red or white but in orange coating

LIGHT: Half sun or more

SOIL: Tolerant of many types

MOISTURE: Moist

SEASONAL ASPECT: Late summer or fall fruits, when formed

ZONE: 1, 2, 3

USES: Many, depending on whether the plant is an evergreen shrub, an evergreen climber or a deciduous shrub or tree

CARE: Guard against Euomymus scale

PROPAGATION BY: Cuttings

NATIVE OF: w. Asia except as noted above

OTHER: Native Deciduous Species:

Strawberry Bush, *E. americanus*; slender, deciduous shrub to 6′ high with green stem and branches; leaves lanceolate, to 2″ long and toothed; flowers green.

Wahoo, *E. atropurpureus*; shrub or small tree to 18′; leaves elliptic, toothed and to 5″ long; flowers purple.

Introduced Evergreen Species:

E. fortunei; China; mostly clinging vines by aerial roots, or ground over; many varieties.

E. japonicus; Japan; erect shrubs; many varieties.

Introduced Deciduous Species:

E. bungeanus; China, Manchuria; shrub or small tree; leaves abruptly long pointed, petioles to 1″ or more long.

E. europaeus, Eu., w. Asia; large shrub; leaves short pointed; petioles less than 1/2″ long.

Evodia

Evodia daniellii(Euodia)

Rue Family

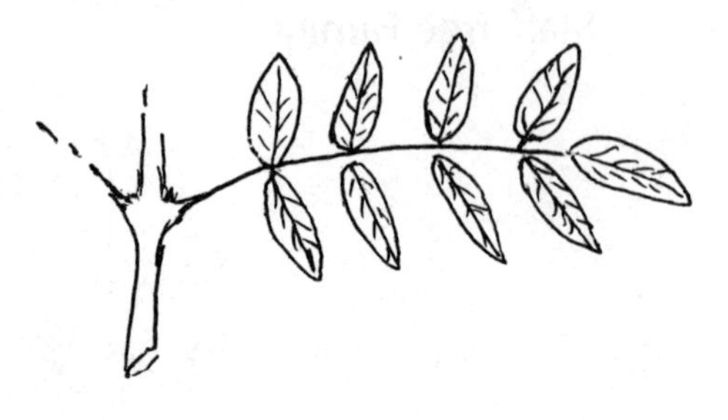

GROWTH HABIT: Open branched, deciduous tree to 25′; rapid grower
LEAVES: Opposite, pinate, 8-12″ long with 7-11 ovate, finely-toothed leaflets
FLOWERS: Whitish, small, in rather large flattish clusters
FRUITS: Lusterous black and less than 1/4″ wide
SEEDS: Few
LIGHT: Full sun
SOIL: Any type
MOISTURE: Moist
SEASONAL ASPECT: Late summer flowers
ZONE: 1, 2, 3
USES: Shrub or woodland border, specimen
CARE: None
PROPAGATION BY: Cuttings, seedlings
NATIVE OF: China to Korea

False-Cypress

Chamaecyparis spp.

Cypress Family

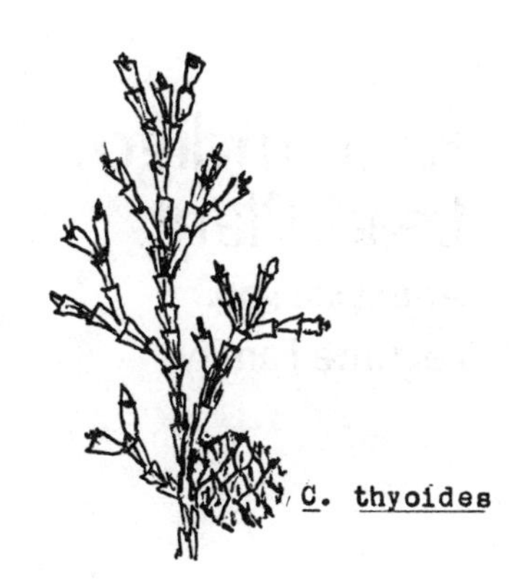

GROWTH HABIT: Evergreen trees with one main trunk and slender branches, branchlets flattened
LEAVES: Scale-like and opposite; needle-like in seedlings
FLOWERS: Male inconspicuous as very small yellow catkins; female as globose cones 1/4 - 1/2″ wide
FRUITS: Small woody cones
SEEDS: Winged
LIGHT: Partial or full sun
SOIL: Clay loam with humus
MOISTURE: Moist to wet
SEASONAL ASPECT: None
ZONE: 1, 2, 3
USES: Freestanding, background, framing
CARE: None
PROPAGATION BY: Seedlings and cuttings
NATIVE OF: See below
OTHER: *C. obtusa,* Hinoki Cypress, Japan; many varieties varying in color and habit of growth.
C. pisifera, Sawara Cypress, Japan; many varieties.
C. thyoides, Atlantic White Cedar; s.e. United States; use along wet margins.

False Indigo, Lead-Plant

Amorpha spp.

Legume Family

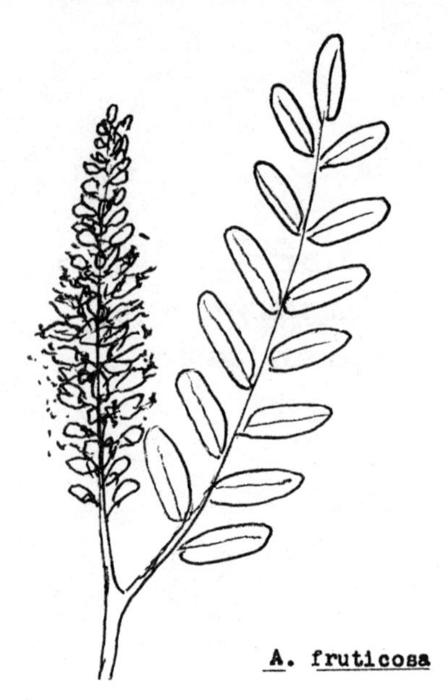

GROWTH HABIT: Deciduous shrubs with feathery foliage
LEAVES: Alternate, pinnate, 11-45 leaflets, oval to elliptic, 1/2 to 1 1/2″ long
FLOWERS: Small, reduced to 1 petal, blue-violet to whitish in dense spikes to 6″ long
FRUITS: Short somewhat flattened and roughened pods
SEEDS: 2
LIGHT: Full or partial sun
SOIL: Tolerant
MOISTURE: Moist to dry
SEASONAL ASPECT: Early summer flowers
ZONE: 1, 2, 3
USES: Shrub, border, margin, rockery, bank
CARE: None
PROPAGATION BY: Seedlings
NATIVE OF: e. North America
OTHER: False Indigo, *A. fruticosa*; nearly glabrous, to 15′ tall; lowest leaflets some distance from stem.
Lead-Plant, *A. canescens*; grayish pubescent, to 4′ tall; lowest leaflets close to stem.

False-Spirea

Sorbaria spp.

Rose Family

GROWTH HABIT: Deciduous shrubs to 8′ high
LEAVES: Alternate, pinnate, with from 13-21 toothed-leaflets
FLOWERS: Small, white and in large terminal inflorescence
FRUITS: Very small pods, up to five per flowers
SEEDS: Few and tiny
LIGHT: Full or partial sun
SOIL: Fertile sandy loam
MOISTURE: Moist to dry
SEASONAL ASPECT: Large inflorescences in late spring
ZONE: 1, 2, 3
USES: Occasional flowering shrub, freestanding
CARE: Some pruning
PROPAGATION BY: Late spring or early summer cuttings
NATIVE OF: Asia
OTHER: S. *aitchisonii*; w. Asia; leaflets over 1/2″ wide; young parts of plant pubescent.

S. *sorbifolia*; e. Asia; leaflets less than 1/2″ wide; young parts of plant glabrous; plant smaller than above.

Fame Flower

Talinum calycinum

Purslane Family

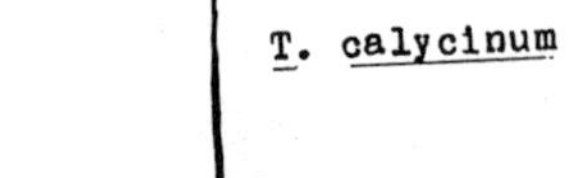

GROWTH HABIT: Fleshy evergreen perennial to 10″ high, from thick root
LEAVES: Clustered at base of stem, terete, to 2″ long
FLOWERS: Raised, rose and 1/2″ wide
FRUITS: 3-valved capsule
SEEDS: Many
LIGHT: Full sun
SOIL: Thin layer of rich organic soil
MOISTURE: Wet to dry
SEASONAL ASPECT: Flowers in summer
ZONE: 1, 2, 3
USES: Rockery
CARE: None
PROPAGATION BY: Seeds or transplants
NATIVE OF: central United States and Mexico
OTHER: *T. teretifolium*; e. United states; very similar.
Jewels-of-Ophir; *T. paniculatum*; erect perennial to 4′ high or more with elliptic leaves to 3″ long and extremely small pink flowers.

Fatshedera

Fatshedera X

Ginseng Family

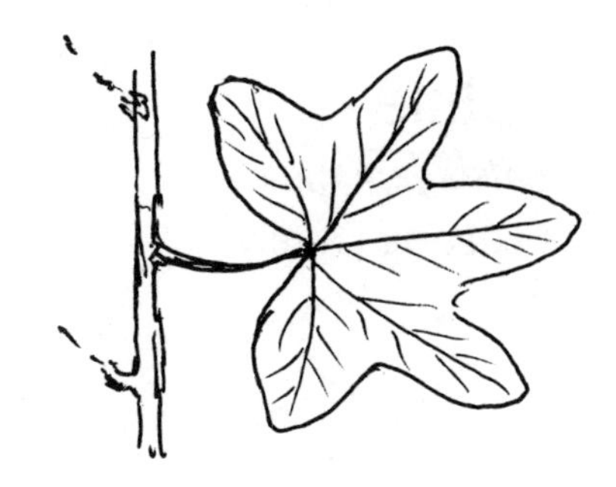

GROWTH HABIT: Evergreen that may be kept as shrub or vine. As a vine it does not twine, have tendrils or aerial roots, therefore must be supported.
LEAVES: Alternate, glossy, 3-5 prominent lobes and to 6″ across
FLOWERS: —
FRUITS: —
SEEDS: —
LIGHT: Half sun or less
SOIL: Fertile of most any type
MOISTURE: Moist
SEASONAL ASPECT: None
ZONE: 1, 2, 3
USES: As a low shrub, or as a vine for ground cover or wherever there may be support
CARE: Some pruning or twining
PROPAGATION BY: Cuttings
NATIVE OF: —
OTHER: This popular hybrid is a cross between Fatsia and Hedera.

Fatsia

Fatsia japonica
Ginseng Family

GROWTH HABIT: Large, bushy, evergreen shrub, essentially a foliage plant
LEAVES: Orbicular in outline, to a foot across, cut below the middle into 5-9 toothed lobes; petioles to a foot long
FLOWERS: Small, 5-parted, whitish borne in many umbels in a large inflorescence in summer or early fall
FRUITS: Dark, 1/4″ wide, globular, topped by two persistent styles
SEEDS: Few
LIGHT: Full shade
SOIL: Tolerant to most types
MOISTURE: Moist
SEASONAL ASPECT: None
ZONE: 1, 2
USES: Effective for winter greenery or to give tropical effect
CARE: Some types may require occasional pruning to keep compact
PROPAGATION BY: Seedlings
NATIVE OF: Japan
OTHER: This attractive foliage plant will add distinction.

Christmas Fern

Polystichum achrosticoides

Fern Family

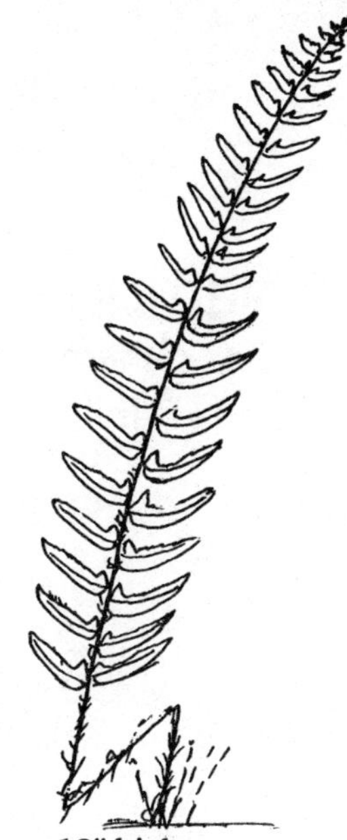

GROWTH HABIT: Dark evergreen fern to 18″ high
LEAVES: One-pinnate, lusterous, clustered
FLOWERS: No true flowers
FRUITS: Dark brown, sporangia form on underside of frond toward the tip
SEEDS: Spores
LIGHT: Shade, or a little sun
SOIL: Rich loam with lots of humus
MOISTURE: Moist
SEASONAL ASPECT: None
ZONE: 1, 2, 3
USES: Interest plants, where conditions allow
CARE: None
PROPAGATION BY: Young plants and transplants
NATIVE OF: e. United States

Cinnamon Fern

Osmunda cinnamomea

Osmunda Family

GROWTH HABIT: Large, attractive, deep-rooted, deciduous fern
LEAVES: One-pinnate, to 30″ long and several per crown
FLOWERS: Fertile fronds about the height of sterile ones and cinnamon in color; no true flowers
FRUITS: Fertile frond consists of sporangia
SEEDS: Spores
LIGHT: Full sun for best development
SOIL: Silt or clay
MOISTURE: Moist to wet
SEASONAL ASPECT: Late spring when cinnamon-brown stems and sporangia are mature
ZONE: 1, 2, 3
USES: Interest plants in or around wet spots or margins
CARE: None
PROPAGATION BY: Root crowns
NATIVE OF: s.e. United States

Holly Fern

Cyrtomium falcatum
(C. imbricatum)

Fern Family

GROWTH HABIT: An evergreen fern reaching 2′ in height
LEAVES: Pinnate, leaflets alternate, toothed or toothless and glossy green
FLOWERS: A spore plant, no flowers
FRUITS: Spore cases in clusters on back of leaf
SEEDS: Spores
LIGHT: Shade
SOIL: Fertile loam
MOISTURE: Moist
SEASONAL ASPECT: Winter greenery
ZONE: 1, 2
USES: As a lower evergreen, accent
CARE: Plant where trees or shrubs protect against cold
PROPAGATION BY: Clump division
NATIVE OF: Asia, s. Africa, Polynesia

Resurrection Fern

Polypodium polypodioides

Fern Family

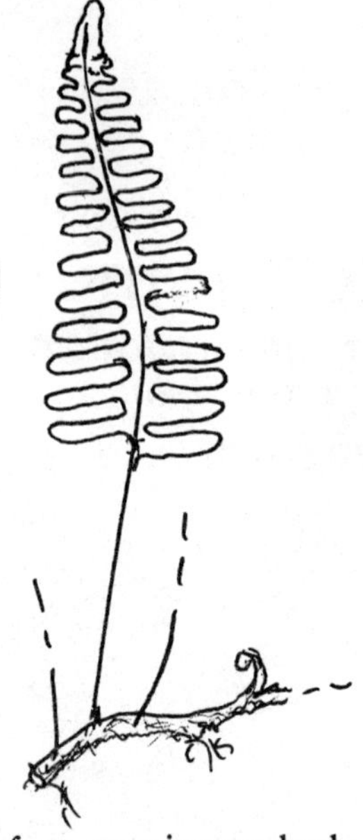

GROWTH HABIT: Small evergreen fern growing on bark of trunks and larger branches of deciduous trees; colony forming by slender rhizomes
LEAVES: One-pinnate with entire, linear segments; to 6″ long
FLOWERS: No true flowers
FRUITS: None as such; spore cases clustered as brown dots on back side of leaf
SEEDS: Spores
LIGHT: Much shade
SOIL: —
MOISTURE: Moist
SEASONAL ASPECT: Much withered and twisted in dry weather it quickly resurrects in wet weather and shows up green
ZONE: 1, 2, 3
USES: Interest
CARE: None
PROPAGATION BY: Colony transplants, spores
NATIVE OF: s.e. United States

Elder

Sambucus canadensis

Honeysuckle Family

GROWTH HABIT: Rather coarse deciduous shrub with several little-branched stems which have white pith equal to at least half their diameters
LEAVES: Opposite, pinnate, 5-11 sharply toothed leaflets
FLOWERS: Small white flowers borne in very large flat terminal clusters
FRUITS: Small blackish berries
SEEDS: Few per berry
LIGHT: Full or partial sun
SOIL: Most types
MOISTURE: Very moist to fairly dry
SEASONAL ASPECT: Flowers in mid-summer, fruits in late summer
ZONE: 1, 2, 3
USES: Bank, low margin - use only in large area
CARE: Renewal pruning occasionally
PROPAGATION BY: Seeds and stolons
NATIVE OF: e. United States
OTHER: This species exists in a number of varieties and although other species are available they offer little difference.

Fetter-Bush

Eubotrys racemosa

Heath Family

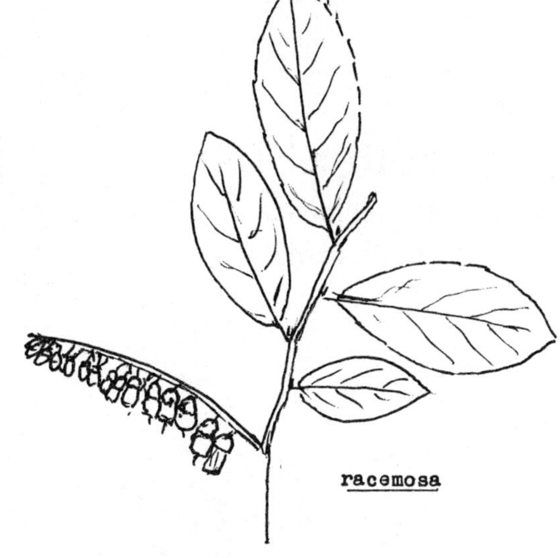

GROWTH HABIT: Deciduous shrub to 7′ high
LEAVES: Alternate, elliptic, to 3 1/2″ long, much smaller back from twig tip
FLOWERS: Racemes of small waxy white or pinkish, urn-shaped and 1/4″ long
FRUITS: Globose capsule
SEEDS: Several, somewhat crescent shaped
LIGHT: Mostly shade
SOIL: Rich, acid
MOISTURE: Moist to wet
SEASONAL ASPECT: Racemes of white flowers in spring; scarlet leaves in fall
ZONE: 1, 2
USES: Low or wet place
CARE: None
PROPAGATION BY: Seedlings or cuttings
NATIVE OF: e. United States
OTHER: *E. recurva*; a similar but somewhat larger species is native to the mountain region.

Fetter-Bush

Lyonia lucida

Heath Family

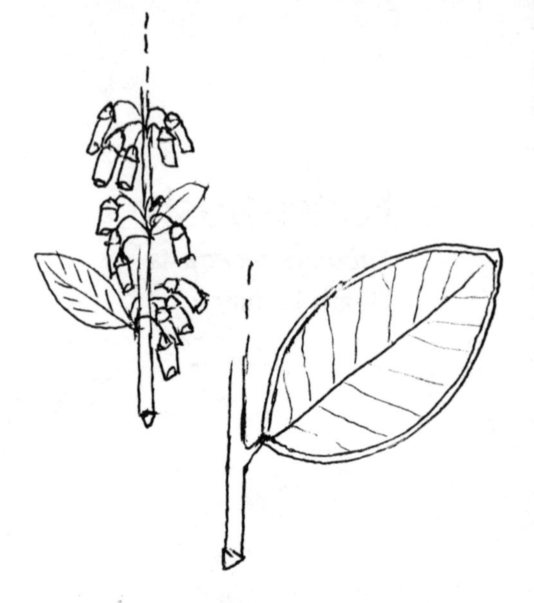

GROWTH HABIT: Weak or slender, colony forming evergreen shrub to 7′ high
LEAVES: Alternate, elliptic, glossy, entire, to 3″ long and with distinct vein very near and parallel to margin
FLOWERS: White to pinkish, tubular, 1/4″ long and in axillary clusters
FRUITS: Short capsule
SEEDS: Very small
LIGHT: At least partial shade
SOIL: Acid, alluvium, silty or sandy loam
MOISTURE: Wet to moist and acid
SEASONAL ASPECT: Spring flowers
ZONE: 1, 2
USES: Low or wet bank, or margin
CARE: None
PROPAGATION BY: Root divisions, seedlings
NATIVE OF: s.e. United States

Fetter-Bush

Pieris floribunda

Heath Family

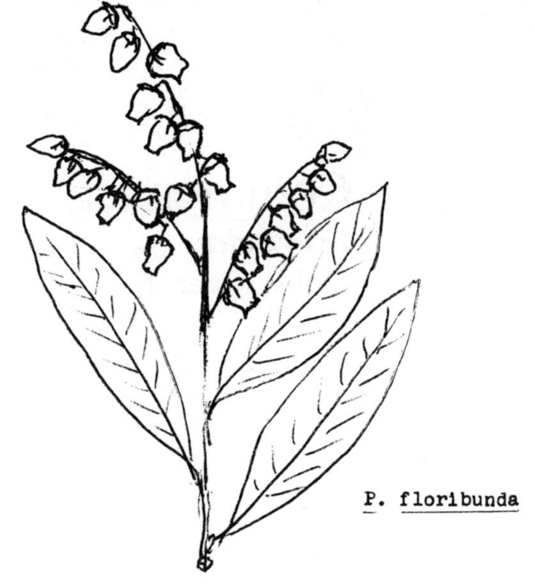

GROWTH HABIT: Erect, freely branched evergreen shrub to 5′ high
LEAVES: Alternate, lanceolate to oblong, entire or nearly so, to 2 1/2″ long
FLOWERS: White, urn-shaped, 1/4″ long and in racemes, erect and clustered at branch tips; flower buds formed in autumn
FRUITS: Short, woody capsule
SEEDS: Several, light brown and small
LIGHT: Partial or full sun
SOIL: Thin, shallow clay loam
MOISTURE: Moist
SEASONAL ASPECT: Spring flowers
ZONE: 3
USES: Specimen
CARE: None
PROPAGATION BY: Seedlings
NATIVE OF: e. United States
OTHER: *P. japonicum*; similar, flowers very early; flower clusters drooping (sometimes listed as *Andromeda*).

Common Fig

Ficus carica

Mulberry Family

GROWTH HABIT: Much branched, deciduous shrub or small tree
LEAVES: Alternate, thickish, broadly ovate to orbicular, usually deeply 3-5 lobed and to 8″ long
FLOWERS: Tiny and inside a closed fleshy structure
FRUITS: Somewhat pear-shaped and variable in size as to variety, 1 - 2 1/2″ long
SEEDS: Small and many
LIGHT: Full or lots of sun
SOIL: Tolerant to many types
MOISTURE: Moist
SEASONAL ASPECT: None
ZONE: 1, 2
USES: Shade, background shrub, fruit
CARE: Freezing may cause die-back but the roots send up sprouts in the spring
PROPAGATION BY: Cuttings
NATIVE OF: Asia Minor
OTHER: Many varieties and several other species are cultivated.

Fig-Vine, Creeping-Fig

Ficus pumila

Mulberry Family

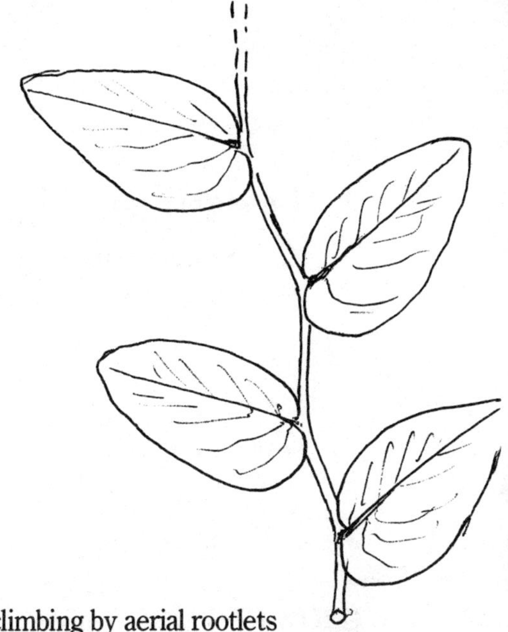

GROWTH HABIT: Evergreen vine climbing by aerial rootlets
LEAVES: Alternate, dark green, 1-1 1/2″ long on vertical or climbing branches 1 1/2 - 3″ long on the horizontal branches
FLOWERS: Rarely formed
FRUITS: —
SEEDS: —
LIGHT: Sun or partial shade
SOIL: Fertile loam
MOISTURE: Moist
SEASONAL ASPECT: None
ZONE: 1, 2
USES: Masonry wall, planter
CARE: Pruning to control extent of coverage desired
PROPAGATION BY: Cuttings
NATIVE OF: e. Asia

Firespike

Odontonema strictum

Acanthus Family

GROWTH HABIT: Semievergreen shrub to 5′
LEAVES: Elliptic, to ovate, 3-6″ long entire and glabrous
FLOWERS: About 1″ long, bright crimson and in terminal spikes; very showy
FRUITS: Capsule
SEEDS: 2 per cell
LIGHT: Full or partial sun
SOIL: Fertile loam
MOISTURE: Moist or dry
SEASONAL ASPECT: Summer flowers
ZONE: 1 (2)
USES: Patio, terrace or protected place
CARE: Protect from cold but in case of die-back it will likely sprout-up again in spring
PROPAGATION BY: Root sprouts, seedlings, cuttings
NATIVE OF: central America

Firethorn

Pyracantha coccinea

Rose Family

GROWTH HABIT: Hardy evergreen shrub with painful branch-tip thorns
LEAVES: Alternate, toothed narrow-elliptic, blunt-tipped and 2″ long
FLOWERS: White, small, in dense clusters and produced abundantly
FRUITS: 1/4″ wide, bright red to orange and produced abundantly
SEEDS: Few
LIGHT: Full or partial sun
SOIL: Tolerant
MOISTURE: Tolerant
SEASONAL ASPECT: Brightly-colored fruits in fall and winter
ZONE: 1, 2, 3
USES: Hedge, freestanding, occasional shrub, background, espalier
CARE: Guard against red spiders
PROPAGATION BY: Seedlings and cuttings
NATIVE OF: s. Europe to w. Asia
OTHER: Varieties of this as well as other species are cultivated; sometimes escaped.

Flowering Quince

Chaenomeles lagenaria

Rose Family

GROWTH HABIT: Stiff-branched, thorny deciduous shrub developing a much-branched top

LEAVES: Alternate, oblong, to 3″ in length, smooth and shining above; stipules usually large

FLOWERS: Numerous, showy to 2″ wide, red to white

FRUITS: 1 - 1 1/2″ long, greenish-yellow, hard and apple-like

SEEDS: Several

LIGHT: Full sun for best blossoms

SOIL: Tolerant

MOISTURE: Moist

SEASONAL ASPECT: Flowers in spring before the leaves

ZONE: 1, 2, 3

USES: Flowering shrub

CARE: Maybe some pruning

PROPAGATION BY: Cuttings

NATIVE OF: China and Japan

OTHER: Several varieties of this and at least two other species are cultivated; all are sometimes confused with *Cydonia*. Dwarf Flowering Quince; *C. japonica*; to 3′ high.

Fog Fruit

Lippia nodiflora (Phyla)

Vervain Family

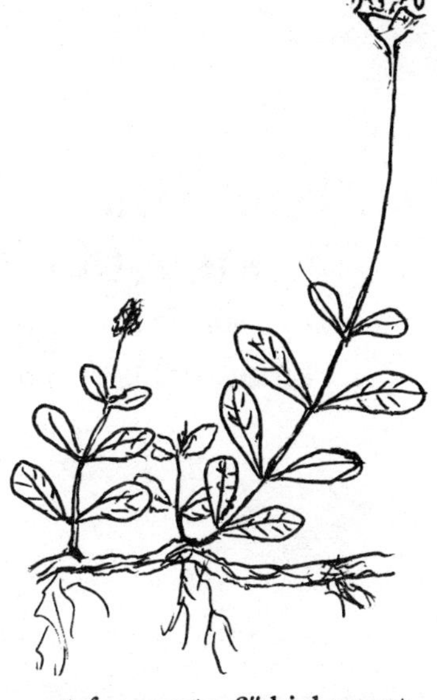

GROWTH HABIT: Prostrate evergreen mat-former to 6″ high; rooting at the nodes
LEAVES: Opposite, toothed, mostly widest toward tips and to about 1″ long
FLOWERS: Lavender, small and in dense heads
FRUITS: Small light brown nutlets
SEEDS: 1 per nutlet
LIGHT: Full sun
SOIL: Sandy or silty loam
MOISTURE: Moist to dry
SEASONAL ASPECT: Summer-long flowers
ZONE: 1 (2)
USES: Ground cover in open places
CARE: None
PROPAGATION BY: Rooted stems
NATIVE OF: s.e. United States

Formosa Honeysuckle

Leycesteria formosa

Honeysuckle Family

GROWTH HABIT: Deciduous shrub to 6′
LEAVES: Opposite, ovate, sometimes toothed, to 5″ long
FLOWERS: Purplish, 1/2″ or more long, in whorls subtended by colored bracts
FRUITS: Berry
SEEDS: Many
LIGHT: Partial shade
SOIL: Most any type
MOISTURE: Moist
SEASONAL ASPECT: Late summer flowers
ZONE: 1, 2, 3
USES: Occasional shrub, interest
CARE: None
PROPAGATION BY: Cuttings or seedlings
NATIVE OF: Himalayas to s. w. China
OTHER: A vigorous grower.

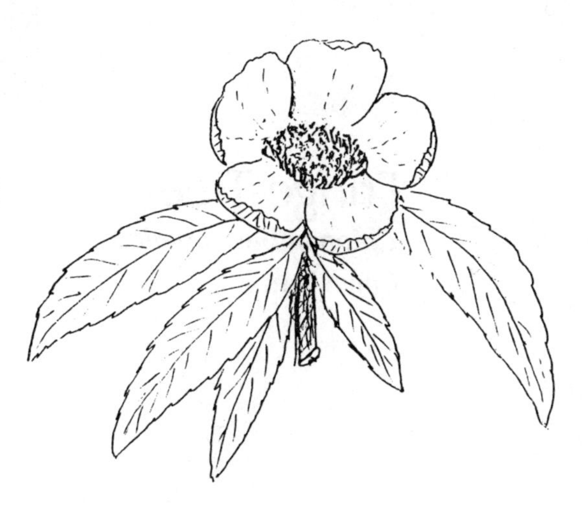

Franklin Tree

Franklinia alatamaha

Tea Family

GROWTH HABIT: Deciduous shrub
LEAVES: Alternate, oblong to ovate, 4-6″ long, toothed toward tips, glossy above, silky beneath
FLOWERS: White, nearly sessile in leaf axils and about 3″ wide
FRUITS: Globose capsule about 1/2″ wide
SEEDS: Angled but not winged
LIGHT: Shifting shade
SOIL: Heavy, as a clay loam
MOISTURE: Moist
SEASONAL ASPECT: Late summer flowers and red leaves in autumn
ZONE: 1, 2
USES: An interest plant, informal shrub border, specimen
CARE: Control competition and keep moist
PROPAGATION BY: Cuttings and seeds
NATIVE OF: Very possibly China
OTHER: The one plant found in the Altamaha River Swamp in south Georgia, and on which the name and description are based, was viewed as being either the first or the last of the species. Suspicion now has it that a hoax was perpetrated by the discoverer, Bartram.

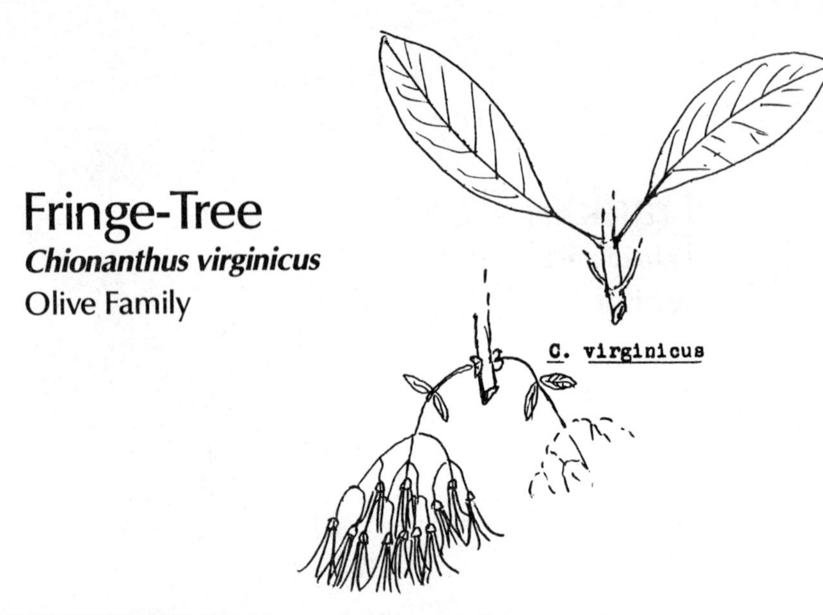

Fringe-Tree

Chionanthus virginicus

Olive Family

GROWTH HABIT: Deciduous shrub or small tree
LEAVES: Subopposite, elliptic, entire, smooth and 3-5″ long
FLOWERS: White with very long, narrow petals, produced in abundance in loose clusters before the leaves appear
FRUITS: Dark blue with thin pulp, 1/2″ long
SEEDS: 1 per drupe
LIGHT: Full sun for best flowers
SOIL: Fertile acid loam with ample humus
MOISTURE: Moist but well-drained
SEASONAL ASPECT: Early spring flowers
ZONE: 1, 2, 3
USES: Freestanding or interest plant with early flowers and fruits throughout autumn
CARE: None
PROPAGATION BY: Seeds or cuttings
NATIVE OF: s.e. United States
OTHER: Chinese Fringetree, *C. retusus*; similar to above. In each of the above, care should be taken to get specimens that flower before the leaves. As in *Forsythia*, some of the effect is lost if leaves occur with the flowers.

Fuchsia

Fuchsia magellanica

Evening-Primrose Family

F. magellanica

GROWTH HABIT: Deciduous subshrub to 3′ high
LEAVES: Opposite or whorled, ovate, finely toothed and about 1″ long
FLOWERS: Red calyx; purplish petals; about 1″ long
FRUITS: Soft berry
SEEDS: Several
LIGHT: Partial sun
SOIL: Fertile and porous
MOISTURE: Moist
SEASONAL ASPECT: Flowers in late spring
ZONE: 1 (2)
USES: Pot, patio, bed, terrace
CARE: Grow in place protected from cold
PROPAGATION BY: Seeds, seedlings
NATIVE OF: Peru, Chile
OTHER: *F. coccinea* and *F. hybrida* similar and sometimes in use.

Galax

Galax aphylla
Diapensia Family

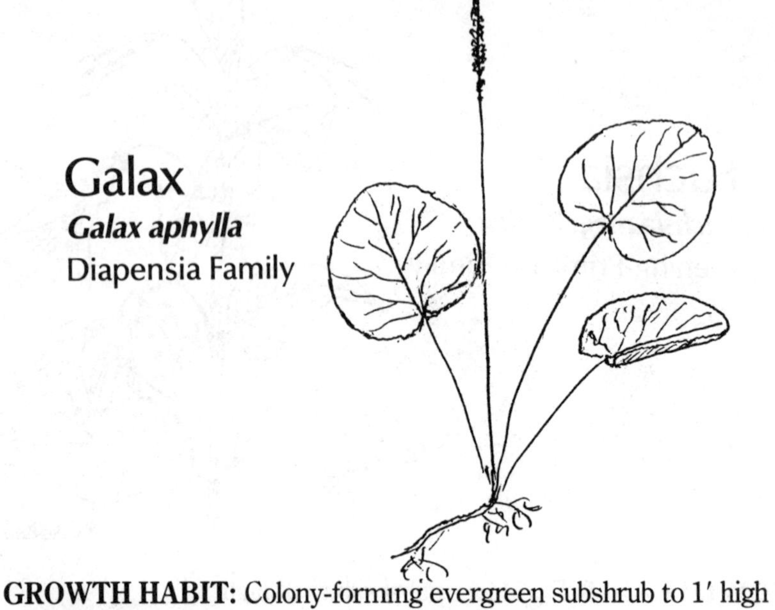

GROWTH HABIT: Colony-forming evergreen subshrub to 1′ high
LEAVES: Roundish, to 3″ wide, with shallow teeth, leathery
FLOWERS: Very small, greenish and numerous on raised flower stalk
FRUITS: Very small capsules
SEEDS: Tiny
LIGHT: Partial shade
SOIL: Fertile loam
MOISTURE: Dry to moist
SEASONAL ASPECT: None
ZONE: (1), 2, 3
USES: Interest, specimen
CARE: None
PROPAGATION BY: Colony division
NATIVE OF: s.e. United States

Germander

Teucrium fruticans

Mint Family

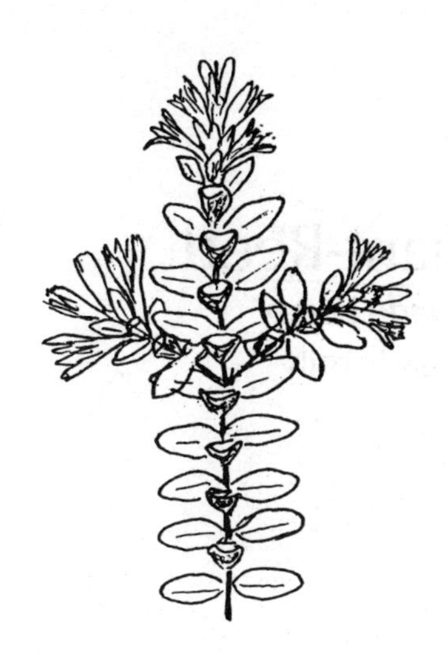

GROWTH HABIT: Evergreen shrub to 3′ high; stem and leaves, especially the younger, covered with gray or yellowish hairs
LEAVES: Opposite, short stalked, round tipped, entire narrowly ovate and 1″ or more long
FLOWERS: Blue, about 1″ long and in short axillary racemes
FRUITS: 4 very small nutlets
SEEDS: 1 per nutlet
LIGHT: Full to three quarters sun
SOIL: Tolerant to most
MOISTURE: Dry
SEASONAL ASPECT: Summer flowers
ZONE: 1, 2, (3)
USES: Bank, border, bed, terrace, patio
CARE: None
PROPAGATION BY: Seeds, sprouts
NATIVE OF: Europe

Giant-Reed

Arundo donax

Grass Family

GROWTH HABIT: Robust, clump forming reed; 6-15′ tall, dying back to ground level in fall
LEAVES: Alternate, linear, tapered, 1-2″ wide and up to 2′ long
FLOWERS: Numerous small spikelets in terminal panicles
FRUITS: Very thin
SEEDS: Infertile
LIGHT: Full sun
SOIL: Tolerant to many types
MOISTURE: Moist
SEASONAL ASPECT: Showy terminal panicles in summer
ZONE: 1, 2, 3
USES: For interest because of its bold habit
CARE: None, as it tends to spread from clump very slowly. Remove dead tops in winter
PROPAGATION BY: Rhizomes
NATIVE OF: Mediterranean regions
OTHER: Forms with foliage striped with white or yellow are available

Ginkgo

Ginkgo biloba

Ginkgo Family

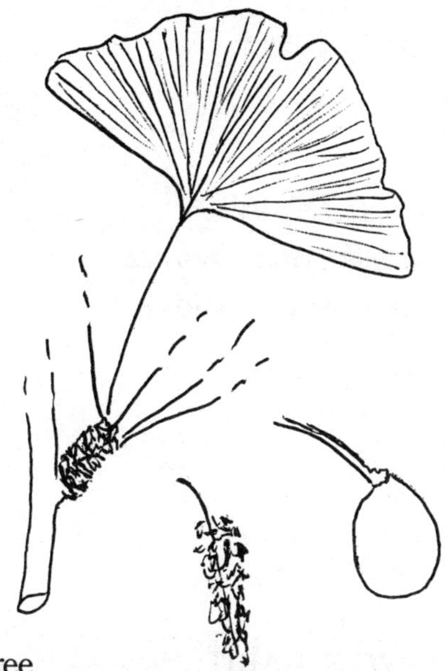

GROWTH HABIT: Deciduous tree
LEAVES: Alternate, fan-shaped and clustered on very short spur twigs
FLOWERS: Male in pediceled catkins; female also pediceled but on separate trees
FRUITS: About 1″ long consisting of yellowish ill-smelling pulp over relative large nut
SEEDS: 1 nut and edible
LIGHT: Full sun
SOIL: Most any loam
MOISTURE: Dry or moist
SEASONAL ASPECT: Leaves turn bright yellow in fall
ZONE: 1, 2
USES: Freestanding, framing
CARE: None
PROPAGATION BY: Seedlings
NATIVE OF: China
OTHER: Because of the ill-smelling fruit pulp, trees producing only the staminate catkins are generally desirable.

Glasswort

Salicornia virginica

Goosefoot Family

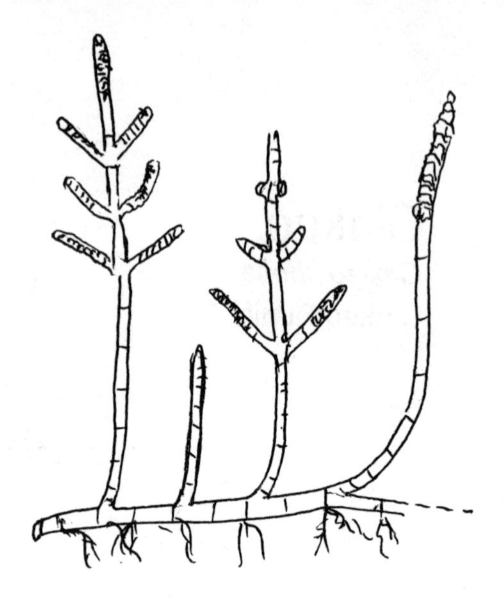

GROWTH HABIT: Ascendant jointed woody-stemmed evergreen perennial, freely branched with thick and fleshy stem; mat forming
LEAVES: Opposite and scale-like
FLOWERS: Barely recognizable along branch tips
FRUITS: Hidden behind fleshy scales
SEEDS: Few and small
LIGHT: Full sun
SOIL: Brackish marsh
MOISTURE: Intertidal
SEASONAL ASPECT: Perennial mat
ZONE: 1, brackish marsh land
USES: Few other plants are as salt tolerant
CARE: None
PROPAGATION BY: Stem sections
NATIVE OF: s.e. United States coast

Glory-Bower

Clerodendron trichotomum

Vervain Family

GROWTH HABIT: Rather coarse deciduous shrub to 10′ or more in height

LEAVES: Opposite, large, ovate, to 6″ long, entire or closely toothed, and pubescent

FLOWERS: White, somewhat fragrant, 1″ or so long and out of reddish-brown calyx

FRUITS: Bright blue berries 1/4″ long supported by brilliant red star-shaped calyces

SEEDS: Mostly 4

LIGHT: Full sun or a little shade

SOIL: Any fertile type

MOISTURE: Moist

SEASONAL ASPECT: Late summer; fruits are perhaps more attractive than flowers

ZONE: 1, 2, 3

USES: Occasional flowering shrub, informal shrub border, specimen

CARE: None

PROPAGATION BY: Seedlings

NATIVE OF: Japan

OTHER: Tube-Flower, *C. indicum*; India; erect, shrub-like plant to 8′ with usually 3 leaves per node; leaves oblanceolate to oblong, entire and to 8″ or more long, corolla white, tube-shaped and 4″ long.

C. splendens; slender, woody-stemmed climber producing dense clusters of scarlet to yellow flowers each 1″ or more long; leaves heart-shaped and to 4″ long; flowers in spring and later also.

Golden-Bell

Forsythia suspensa

Olive Family

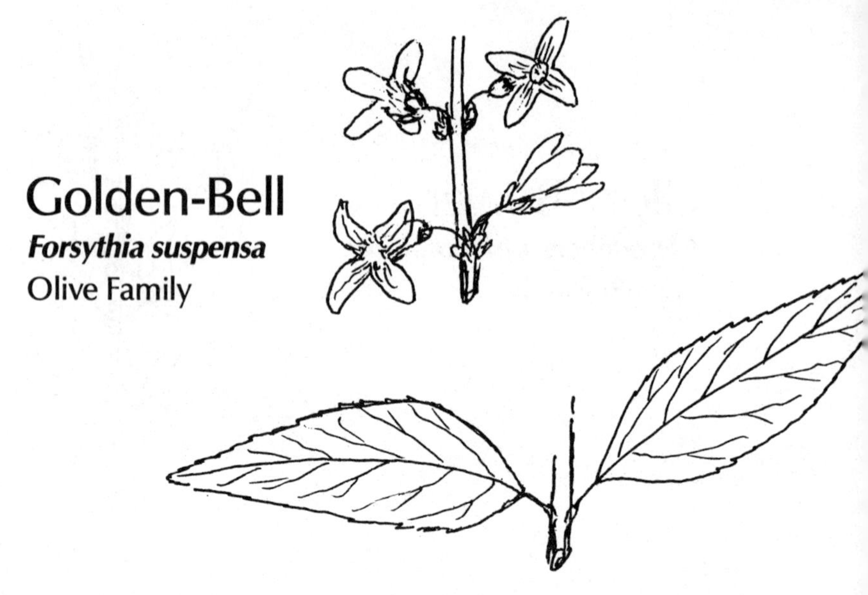

GROWTH HABIT: Deciduous shrub with twigs hollow between the joints

LEAVES: Opposite, simple or sometimes 3-parted, lanceolate and irregularly toothed

FLOWERS: Golden yellow, about 1″ long, 4-parted, most showy of the several species

FRUITS: Woody capsule

SEEDS: Winged

LIGHT: Full sun

SOIL: Fertile loam

MOISTURE: Dry to moist

SEASONAL ASPECT: Spring flowers before or with the leaves

ZONE: 1, 2, 3

USES: Flowering shrub, foundation, unit arrangement

CARE: Some pruning

PROPAGATION BY: Cuttings

NATIVE OF: China

OTHER: Several varieties of this and two or more additional species are grown. The showiest ones produce blossoms ahead of, not with, the leaves.

Goldenchain-Tree

Laburnum anagyroides

Legume Family

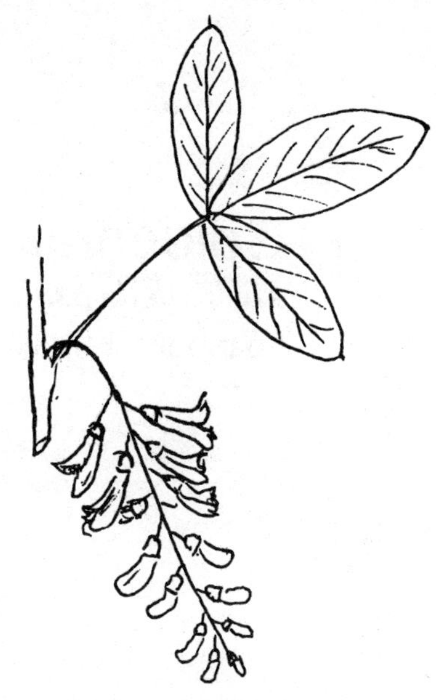

GROWTH HABIT: Wide-spreading, deciduous shrub or small tree
LEAVES: Alternate, 3-foliate, smooth and toothless, leaflets about 2″ long
FLOWERS: About 3/4″ long and in pendant clusters to 1′ long
FRUITS: 2-3″ long pod
SEEDS: Several per pod
LIGHT: Full sun
SOIL: Moderately fertile
MOISTURE: Dry
SEASONAL ASPECT: Early summer flowers
ZONE: 1, (2)
USES: Specimen, for exotic effect
CARE: Some pruning
PROPAGATION BY: Seeds, seedlings
NATIVE OF: s. Europe

Goldenrain-Tree

Koelreuteria paniculata

Soapberry Family

GROWTH HABIT: Small deciduous tree with hemispheric crown
LEAVES: Alternate, odd-pinnate, 7-15 toothed or lobed leaflets
FLOWERS: Bright yellow, numerous and in large panicles
FRUITS: Three-celled, inflated papery capsule
SEEDS: Three, brown
LIGHT: Full sun
SOIL: Tolerant of most
MOISTURE: Moist and well drained to dry
SEASONAL ASPECT: Flowers in late summer
ZONE: 1, 2
USES: Freestanding, street planting
CARE: Very little
PROPAGATION BY: Seeds and seedlings
NATIVE OF: e. Asia

Gopher-Apple

Chrysobalanus oblongifolius (Geobalanus)

Rose Family

GROWTH HABIT: Low evergreen shrub to 1 1/2′ high, from thick underground stem
LEAVES: Alternate, broader toward tip, entire and to 4″ long
FLOWERS: White and in a terminal inflorescence
FRUITS: An inch long drupe
SEEDS: 1
LIGHT: Partial shade
SOIL: Sandy
MOISTURE: Dry
SEASONAL ASPECT: None
ZONE: 1
USES: Specimen
CARE: None
PROPAGATION BY: Seeds and transplants
NATIVE OF: Lower coastal plain, S.C. - Miss.

Greenbrier

Smilax smallii

Lily Family

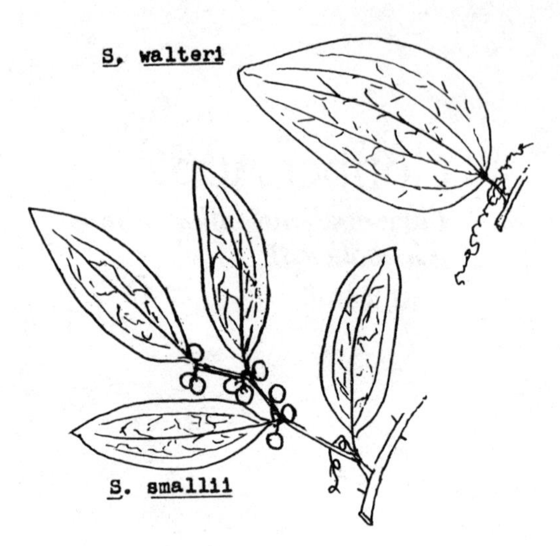

GROWTH HABIT: High-climbing, woody evergreen vine with green, usually spiny stem, and tendrils
LEAVES: Thin, leathery, entire and sometimes mottled with lighter green, shiny above and to 3″ long
FLOWERS: Small, greenish and in axillary clusters
FRUITS: 1/4″ wide black berries
SEEDS: 1-3
LIGHT: Partial sun
SOIL: Almost any
MOISTURE: Moist
SEASONAL ASPECT: None
ZONE: 1, 2
USES: Fence, arbor, trellis, tree
CARE: None
PROPAGATION BY: Seeds or root tubers
NATIVE OF: s.e. United States
OTHER: Fresh leaves desirable for indoor decoration. Red-berried Smilax; *S. walteri*; similar but mostly deciduous and nearly spineless, producing bright red berries in abundance. Requires a very wet site.

Green Fly Orchid

Epidendrum conopseum

Orchid Family

GROWTH HABIT: Epiphytic evergreen growing in crotches and on horizontal branches of old deciduous trees in swamps and other humid situations
LEAVES: Narrowly oblong and to 3″ long
FLOWERS: Grayish or purplish-green, about 1/2″ long and fragrant
FRUITS: 1/2″ long capsule
SEEDS: Many and dustlike
LIGHT: Shade
SOIL: Tree bark
MOISTURE: Humid air
SEASONAL ASPECT: None
ZONE: 1
USES: Specimen
CARE: None
PROPAGATION BY: Transplanting portions of colony
NATIVE OF: Coastal areas, South Carolina - Mississippi

Groundsel, Silverling, Sea-Myrtle

Baccharis halimifolia

Composite Family

GROWTH HABIT: Evergreen shrub to 10′ or more high
LEAVES: Alternate, elliptic to obovate, to 2 1/2″ long, coarsely toothed and whitened
FLOWERS: Male and female on different plants; flowers very small and crowded into heads; cream colored
FRUITS: Achene with silky hairs
SEEDS: 1
LIGHT: Full sun for best development
SOIL: Most any
MOISTURE: Moist
SEASONAL ASPECT: Fall when the silky-haired achenes are mature
ZONE: 1, 2 (3)
USES: In coastal landscaping and elsewhere as shrub borders or as specimen or interest plant
CARE: None
PROPAGATION BY: Seedlings
NATIVE OF: s.e. United States
OTHER: *B. glomeruliflora*; s.e. United States; similar but with heads in clusters of 3′s.
B. angustifolia; s.e. United States; smaller plant with narrow leaves are often mistaken for flowers.

Hackberry

Celtis laevigata

Elm Family

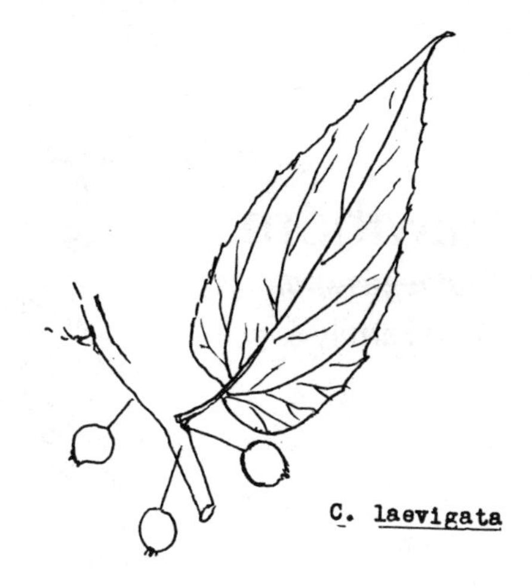

GROWTH HABIT: Medium to large deciduous tree with light gray, somewhat warty bark
LEAVES: Alternate, ovate with long tapering tips, 2-4″ long and yellow-green
FLOWERS: Small greenish and produced on new growth along with the developing leaves
FRUITS: Thin, dark-skinned pulp over a boney stone, 1/4″ wide
SEEDS: 1 per drupe
LIGHT: Full sun or partial shade
SOIL: Tolerant to most, after establishment
MOISTURE: Very tolerant
SEASONAL ASPECT: None
ZONE: 1, 2, 3
USES: Single planting, along streets or wherever a large tree and thin shade are desired
CARE: None
PROPAGATION BY: Seedlings
NATIVE OF: s.e. United States
OTHER: A fast-growing and extremely hardy tree. *C. occidentalis*; native, similar but with dark green foliage, also casting a thin shade.

Hawthorn

Crataegus spp.

Rose Family

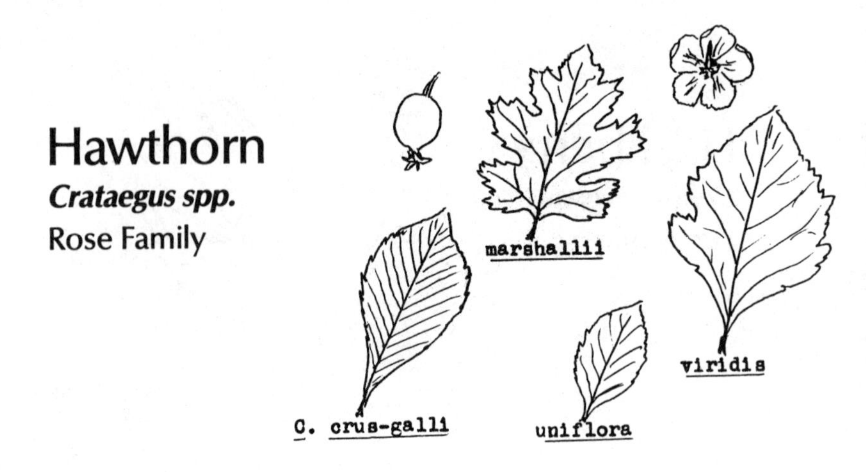

GROWTH HABIT: Deciduous large shrubs or small trees usually with long slender unbranched thorns.

LEAVES: Alternate, variously lobed and toothed

FLOWERS: White or tinted; about 1/2″ wide, and produced abundantly at tips of short twigs

FRUITS: 1/4 - 1/2″ wide and apple-like

SEEDS: 1-5, relatively large and long

LIGHT: Some shade

SOIL: Tolerant

MOISTURE: Moist or dry

SEASONAL ASPECT: Spring flowers

ZONE: 1, 2, 3

USES: Single planting or hedge

CARE: None, except for hedge

PROPAGATION BY: Seedlings and cuttings

NATIVE OF: e. North America, Europe and Asia

OTHER: A number of species are sometimes used as ornamentals but none widely.

C. *crus-galli*; e. United States; small spreading tree; leaves thickish, dull green, obovate and toothed; flowers clustered; use in dry, open sites.

C. *marshallii*; e. United States; small tree or shrub; leaves thin, shining, deeply lobed and toothed; use in moist shaded site.

C. *uniflora*; e. United States; to 8′ high; leaves obovate and toothed; flowers solitary; use in dry sites.

C. *viridis*; e. United States; small tree; leaves toothed, sometimes lobed; fruits small, red, long persistent; use in wet sites.

Hazelnut

Corylus americana

Birch Family

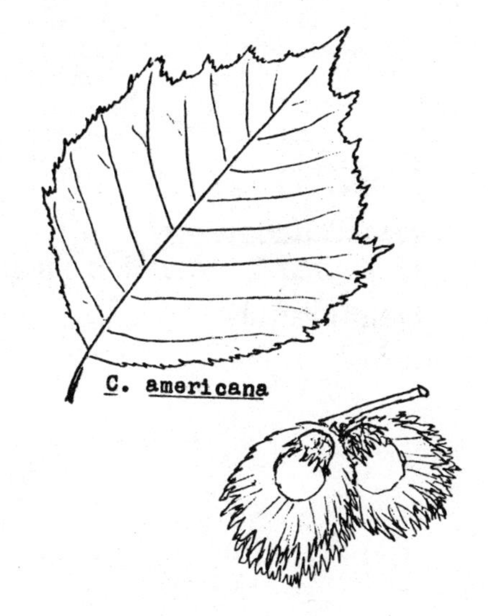

GROWTH HABIT: Deciduous shrub to 8′ high
LEAVES: 2-4″ long, ovate, doubly toothed
FLOWERS: Before the leaves, male as pendulous catkins; female small and hidden by green bracts
FRUITS: 1/2″ long nut partially enclosed by somewhat leafy bracts
SEEDS: Large and edible
LIGHT: Partial sun
SOIL: Fertile
MOISTURE: Moist
SEASONAL ASPECT: None
ZONE: 2, 3
USES: Interest, specimen, informal shrub border, nuts
CARE: None
PROPAGATION BY: Seeds or seedlings
NATIVE OF: e. United States
OTHER: Beaked Hazelnut, *C. cornuta*; similar plant restricted to the mountains and Piedmont.
C. avellana var. contorta; a shrub of interesting growth habit, a variety of Filbert.

Heath

Erica X darleyensis
(E. carnea X E. mediterranea)
Heath Family

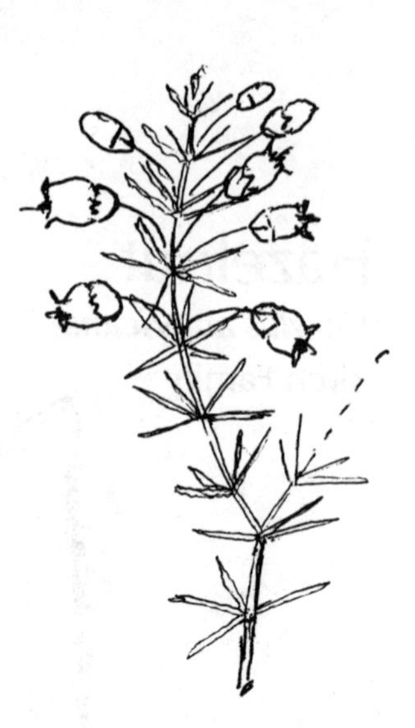

GROWTH HABIT: Evergreen shrub with fine textured foliage, to 2′ high
LEAVES: In whorls of 4 or 5
FLOWERS: Small and pale lilac-pink
FRUITS: Capsule
SEEDS: Many and minute
LIGHT: Sun or some shade
SOIL: Most any acid type
MOISTURE: Moist but well drained
SEASONAL ASPECT: Late fall to early spring flowers
ZONE: 2, 3
USES: Bed, border, edging, rockery
CARE: Acid sand with leaf mold as mulch
PROPAGATION BY: Seedlings or cuttings
NATIVE OF: Europe
OTHER: A vigorous grower, likely to succeed where other species fail.

Hebe

Hebe brachysiphon (Veronica)

Figwort Family

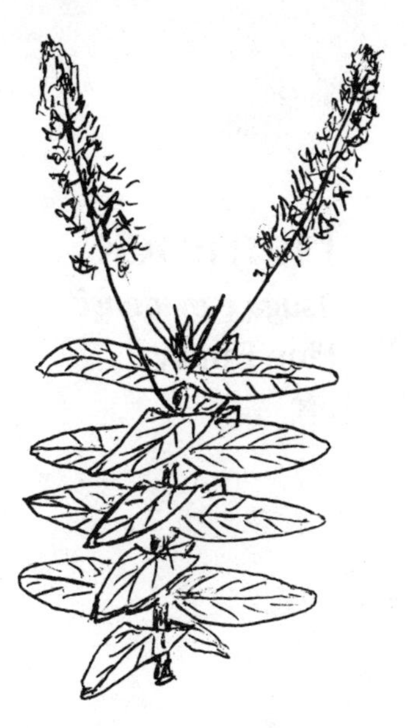

GROWTH HABIT: Dense evergreen shrub forming round bush to 5′ high
LEAVES: Elliptic, to 1″ long and dark green
FLOWERS: Small, white and in 2″ long spikes
FRUITS: Capsule
SEEDS: Several
LIGHT: Full or partial sun
SOIL: Fertile
MOISTURE: Moist
SEASONAL ASPECT: Summer flowers
ZONE: 1, 2, 3
USES: Occasional shrub
CARE: None
PROPAGATION BY: Seeds or cuttings
NATIVE OF: New Zealand
OTHER: Easily mistaken for *H. traversii.*

Hemlock

Tsuga canadensis

Pine Family

GROWTH HABIT: Evergreen tree with horizontal branches
LEAVES: About 1/2″ long, flat, mostly 2-ranked, shiny dark green above and pale beneath because of two whitish lines
FLOWERS: Male inconspicuous; female as cones
FRUITS: Brown ovoid cones about 1/2″ long
SEEDS: Winged
LIGHT: Full or partial sun
SOIL: Clay loam
MOISTURE: Moist
SEASONAL ASPECT: None
ZONE: 3
USES: Freestanding, framing or background
CARE: None
PROPAGATION BY: Seedlings
NATIVE OF: Alabama northward along the mountains
OTHER: *T. caroliniana*; Carolina Hemlock; smaller tree limited to the mountain regions, Georgia to Virginia

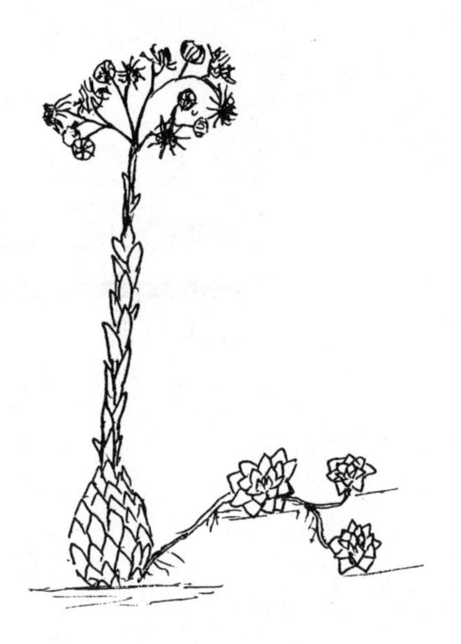

Hens-and-Chickens

Sempervivum tectorum

Orpine Family

GROWTH HABIT: Plant succulent, mostly evergreen and condensed so that the mature appearance is a large rosette surrounded by smaller ones
LEAVES: Thick, pointed, crowded and to 3″ long
FLOWERS: Numerous on raised, hairy stalks, pink, regular and usually 12 parted
FRUITS: Several follicles per flower
SEEDS: Few to many and very small
LIGHT: Full or lots of sun
SOIL: Most any
MOISTURE: Dry
SEASONAL ASPECT: None
ZONE: 1, 2, 3
USES: Around tree base or rock
CARE: None
PROPAGATION BY: The "Chickens", the stoloniferous shoots with clustered leaves at tips
NATIVE OF: Europe
OTHER: Other species are available

Hercules Club

Zanthoxylum clava-herculis

Rue Family

GROWTH HABIT: Deciduous, prickly aromatic shrub or small tree; prickles on trunk, branches and leaves
LEAVES: Alternate with 5-17 leaflets, each to 2 1/2″ long, rather ovate and slightly toothed
FLOWERS: Male and female on separate plants, but similar, numerous, terminal and in large clusters
FRUITS: A small, somewhat fleshy pod
SEEDS: 1 and shining black
LIGHT: Full or partial sun
SOIL: Sandy
MOISTURE: Moist
SEASONAL ASPECT: None
ZONE: 1
USES: Barrier, margin, specimen
CARE: None
PROPAGATION BY: Seedlings
NATIVE OF: s.e. United States coastal area
OTHER: Toothache-Tree, Prickly Ash, *Z. americana*; widespread in e. United States, but rare and coastal here.

Hickory

Carya spp.

Walnut Family

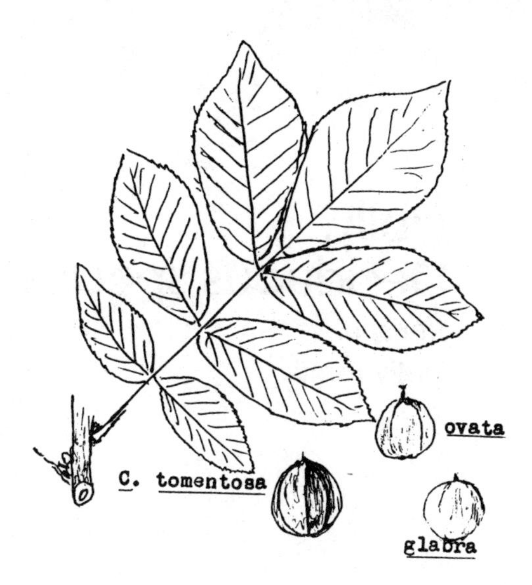

GROWTH HABIT: Large deciduous trees with hard strong wood
LEAVES: Alternate, odd-pinnate with 5-15 toothed leaflets
FLOWERS: Male in pendant 3 branched catkins before the leaves; female without petals and 1-few together
FRUITS: Nut in 4-valved husk
SEEDS: Large and edible
LIGHT: Partial or full sun
SOIL: Fertile loam
MOISTURE: Moist
SEASONAL ASPECT: Bright yellow leaves in fall
ZONE: 1, 2, 3
USES: Background, framing, freestanding, shade, nuts
CARE: None
PROPAGATION BY: Seedlings
NATIVE OF: e. North America
OTHER: White Hickory, *C. tomentosa*; 7-9 leaflets.
Pecan, *C. illinoensis*; 9-15 leaflets.
Shagbark Hickory, *C. ovata*; 5-(7) leaflets.
Pignut, *C. glabra*; leaflets usually 5. Several other species are native to our area and hardy.

Holly, Deciduous species

Ilex spp.

Holly Family

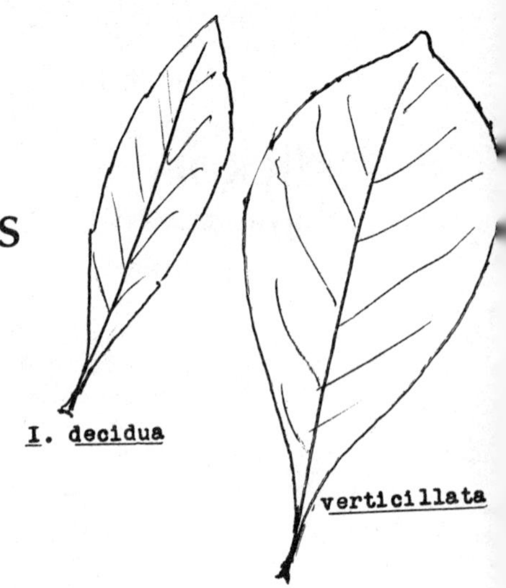

GROWTH HABIT: Deciduous shrubs or small trees
LEAVES: Alternate, thin, toothed, or doubly so and usually pubescent beneath, at least on midrib
FLOWERS: Male and female on separate plants, both greenish, small and inconspicuous
FRUITS: Red berries abundantly produced and persisting well into the winter
SEEDS: Few and bony
LIGHT: Half sun or less
SOIL: Most any type rich in organic matter
MOISTURE: Moist to wet
SEASONAL ASPECT: Bright red (to orange) fruits in winter
ZONE: 1, 2, 3
USES: Accent, specimen
CARE: None
PROPAGATION BY: Cuttings or seedlings
NATIVE OF: e. United States
OTHER: Winterberry, *I. decidua*; to 25′ high; fruit in clusters; leaves mostly n short spur twigs.

Possum-Haw, *I. verticillata*; to 8′ high; fruit not in clusters, no spur twigs; use in wet places.

Holly, Evergreen, nonspiny-leaved species

Ilex spp.

Holly Family

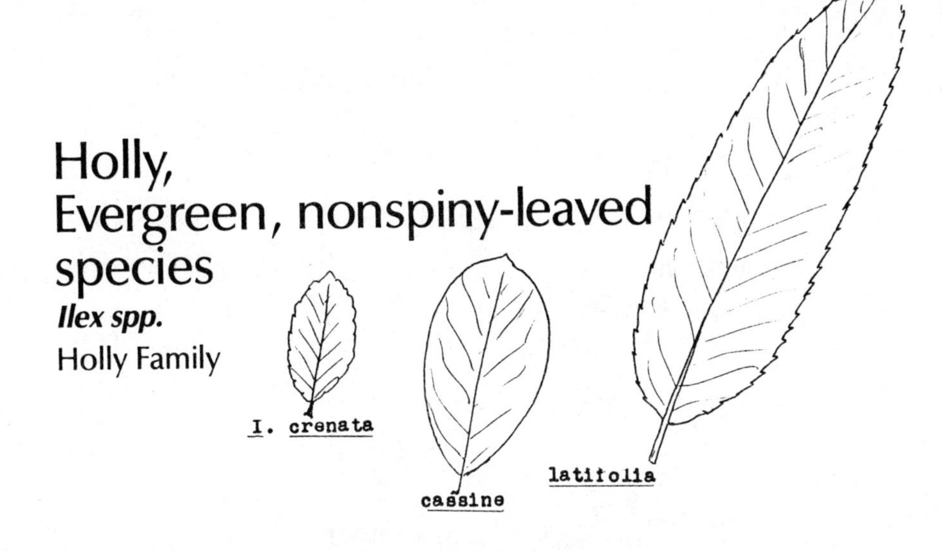

GROWTH HABIT: Evergreen shrubs or small trees
LEAVES: Alternate, linear to oval to obovate, 1-7″, some toothed
FLOWERS: Male and female usually on different plants, both small, greenish, inconspicuous
FRUITS: Black or red berries
SEEDS: Few and bony
LIGHT: Full or half sun
SOIL: Tolerant of most fertile types
MOISTURE: Moist
SEASONAL ASPECT: Fall and winter when berries are ripe
ZONE: 1, 2, 3
USES: Hedge, border, foundation, unit arrangement
CARE: Some pruning
PROPAGATION BY: Cuttings
NATIVE OF: see below
OTHER: *I. crenata*; Japan; black berries, many varieties as *convexa, helleri, latifolia, microphylla, rotundifolia.*

I glabra, e. United States; black berries; leaves usually toothed near apex.

I. cassine; s.e. United States; red berries; leaves entire, 2″- long.

I. latifolia; Japan; red berries; large shrub; leaves 3-8″ long and toothed.

I. myrtifolia; s.e. United States; red berries; leaves entire, 1/2 - 1 1/2″ long.

I. vomitoria; e. United States; red berries; leaves toothed, 1/2 - 1″ long.

myrtifolia
vomitoria
glabra

Holly,
Evergreen, spiny-leaved species

Ilex spp.

Holly Family

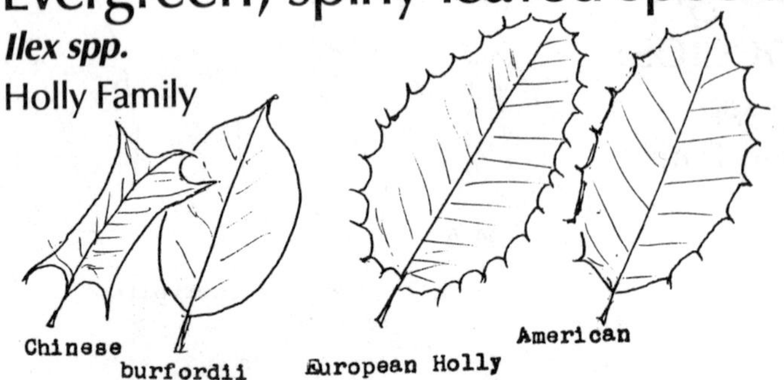

GROWTH HABIT: Evergreen trees or shrubs with spiny leaves and smooth light gray bark

LEAVES: Alternate, stiff-coriaceous with spines along margins and at tip, shining and to 3″ long

FLOWERS: Male and female usually on separate plants; both small, greenish-white and axillary

FRUITS: Red berries 1/4 - 1/2″ long

SEEDS: 2-8 and bony

LIGHT: Full or lots of sun

SOIL: Rich loamy type

MOISTURE: Moist

SEASONAL ASPECT: Scarlet or red berries in fall and winter

ZONE: 1, 2, 3

USES: Trees for background, freestanding; shrubs for hedge, unit arrangement

CARE: Shrub type require some pruning

PROPAGATION BY: Cuttings

NATIVE OF: See below

OTHER: American Holly, *I. opaca*; tree, flowers on this year's growth.

European Holly; *I. aquifolia*; tree, flowers on last year's growth.

Chinese Holly, *I. cornuta*; shrub to 10′, flowers on last year's growth. The variety *burfordii* is popular. All show variations in spininess and fruit color.

Honeycup

Zenobia pulverulenta

Heath Family

GROWTH HABIT: Deciduous shrub to 4′ high and weakly colony-forming

LEAVES: Alternate, oval to broadly elliptic, finely and shallowly toothed, mostly whitened beneath; reddish in autumn

FLOWERS: White, cup-shaped, 1/2″ long

FRUITS: Depressed woody capsules

SEEDS: Several and very small

LIGHT: Partial sun

SOIL: Sandy, acid

MOISTURE: Wet to moist

SEASONAL ASPECT: Summer flowers

ZONE: 1

USES: In ericaceous border, woodland planting, wet margin

CARE: None

PROPAGATION BY: Seedlings or colony division

NATIVE OF: s.e. United States

Honeysuckle

Lonicera spp.

Honeysuckle Family

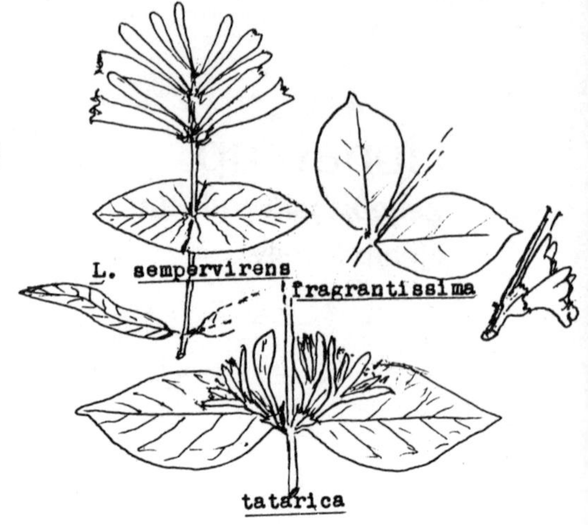

Probably over 100 species and varieties are found in the trade and from the point of view of the landscaper many are similar to the point of being equally useful. Most are vigorous growers, have opposite leaves, show no bright autumn colors, produce white, yellow or red flowers, bear small red or black berries, tolerate most types of soil, love full sun and require some pruning.

Trumpet Honeysuckle, Woodbine, *L. sempervirens*; e. United States; deciduous, woody vine with bright red tubular flowers, some yellow within.

Yellow Honeysuckle; *L. flava*; e. United States; as above but with yellow flowers.

Japanese Honeysuckle; *L. japonica*; Japan; decidous, woody rampant twining vine with white flowers yellowing in age.

Breath-of-Spring. *L. fragrantissima*; e. China; half evergreen bush to 6′ high; creamy-white, very early flowers; very fragrant.

Amur Honeysuckle; *L. maackii*; Asia; deciduous shrub to 15′ high; late white flowers; abundant red fruits.

Tatarian Honeysuckle, *L. tatarica*; s. Rusia; deciduous shrub to 8′ high; pink and white flowers, red berries; vigorous.

Prostrate Honeysuckle, *L. pileata*; China; spreading evergreen shrub with 1/2″ long leaves and fragrant whitish flowers.

Dwarf-leaved Honeysuckle, *L. nitida*; China; evergreen, much branched, densely leafy shrub to 6′; leaves 1/4″ long; cream-colored flowers in spring.

Hop-Tree

Ptelea trifoliata

Rue Family

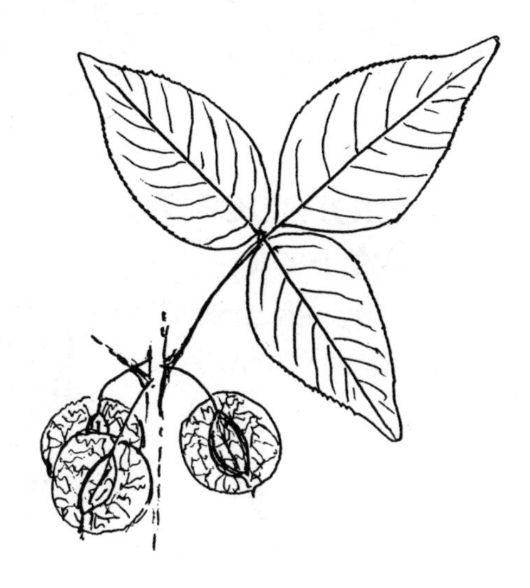

GROWTH HABIT: Deciduous shrub or small tree
LEAVES: Alternate, trifoliate and long petioled; leaflets sessile, more-or-less ovate, entire or barely toothed
FLOWERS: Small, greenish or yellowish-white and clustered
FRUITS: Nutlet surrounded by a thin wing
SEEDS: 1
LIGHT: About half sun
SOIL: Rich loam or sandy loam
MOISTURE: Moist
SEASONAL ASPECT: Late summer fruit
ZONE: 1, 2, 3
USES: Rocky or sandy bank or margin, specimen
CARE: None
PROPAGATION BY: Seeds or seedlings
NATIVE OF: e. United States
OTHER: May be slightly colony forming

Hornbeam, Muscle-Tree

Carpinus caroliniana

Birch Family

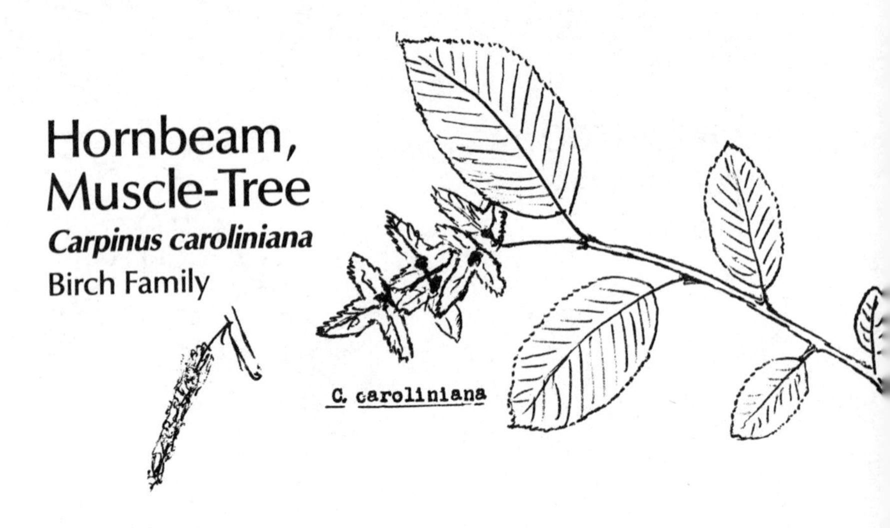

GROWTH HABIT: Small deciduous tree with ridged trunk, gray bark and wide crown

LEAVES: 1-2 1/2″ long, oblong, toothed and pointed

FLOWERS: Male in catkins before the leaves; female in short catkins becoming leafy bracted

FRUITS: 1 or 2 nutlets 1/8″ long per leafy bract

SEEDS: 1 per nutlet

LIGHT: Partial shade

SOIL: Silty loam

MOISTURE: Moist to wet

SEASONAL ASPECT: None

ZONE: 1, 2, 3

USES: Specimen, interest

CARE: None

PROPAGATION BY: Seedlings

NATIVE OF: e. United States

OTHER: A good choice for a low shady corner.

European Hornbean, *C. betulus*; leaves ovate and with 10-14 veins on each side of midvein.

Japanese Hornbean, *C. japonica*; leaves lanceolate and with 20-24 veins on each side of midvein.

Horse-Sugar, Sweetleaf

Symplocos tinctoria

Symplocos Family

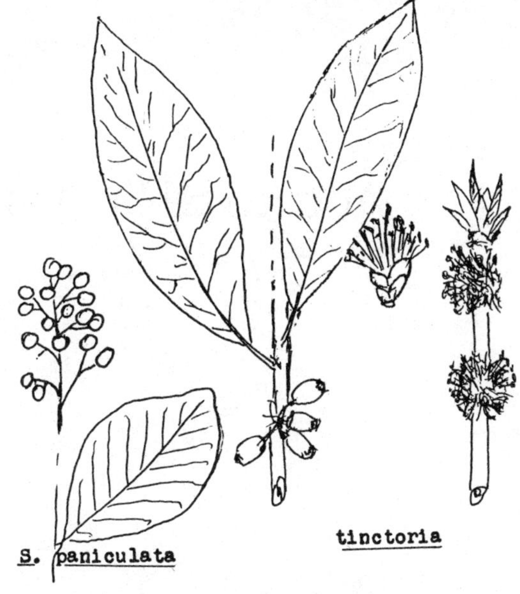

GROWTH HABIT: Well-branched, semievergreen shrub to 12′ high
LEAVES: Alternate, thick, elliptic, ro 4 1/2″ long and somewhat sweet tasting while glandular hairs are beneath
FLOWERS: Yellowish with numerous stamens, fragrant and in dense clusters on the bare stems
FRUITS: Green drupe to 1/2″ long
SEEDS: 1
LIGHT: Partial shade
SOIL: Most any good fertile type
MOISTURE: Moist
SEASONAL ASPECT: Spring flowers before the leaves
ZONE: 1, 2, 3
USES: Occasional shrub, specimen
CARE: None
PROPAGATION BY: Seedlings and cuttings
NATIVE OF: e. United States
OTHER: S. paniculata; Japan, China and the Himalayas; white flowered and blue fruited.

Additional Small Evergreens

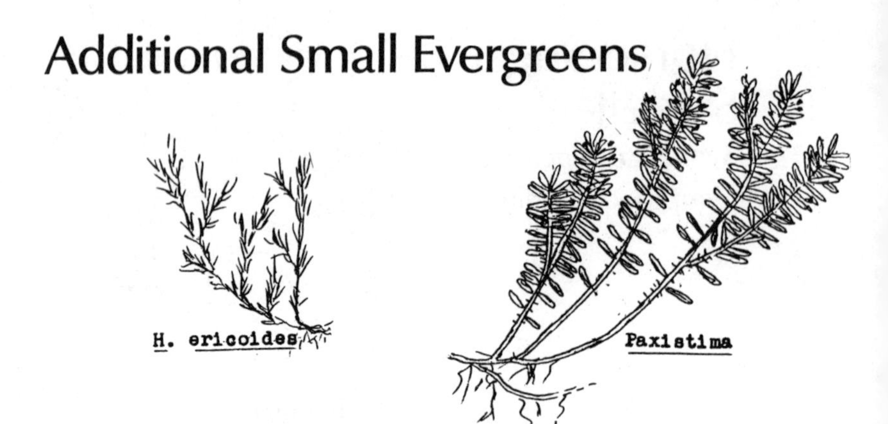

Under appropriate environmental conditions the addition of one or more of the following very small woody evergreens whose southern range limits extend into our area might add interest to any garden or satisfaction to the serious collector.

Hudsonia, Rockrose Family, low or prostrate mat-formers with scale-like or short-pointed leaves and small 5-parted yellow flowers.

H. tomentosa, with closely appressed scale-like leaves; sandy, coastal areas.

H. ericoides, with spreading, pointed leaves; sandy areas, costal plain.

H. montana, similar to the immediate above but limited to high elevations.

Pachistima canbyi, Bittersweet Family, low spreading evergreen shrub with small opposite leaves and small greenish 4-parted flowers; zones 2, 3.

Pyxidanthera barbulata, Pyxie-Moss, Diapensia Family, very small evergreen creeper with pointed leaves and white to pink axillary flowers; sandy, partially shaded areas; zones 1, 2.

Hydrangea, Hills-of-Snow

Hydrangea arborescens var. grandiflora

Saxifrage Family

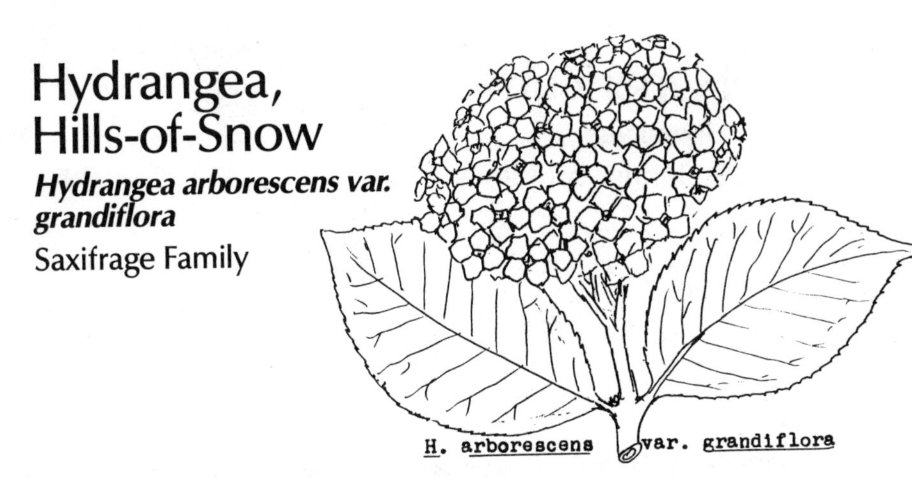

GROWTH HABIT: Somewhat straggling deciduous shrub to 6′ or more high
LEAVES: Opposite, ovate, toothed, paler beneath
FLOWERS: White, blue or pink and in dense rounded or conical clusters, sterile
FRUITS: Small papery capsules on other species
SEEDS: Many and very small on other species
LIGHT: Some shade
SOIL: Fertile, acid soil usually produces blue flowers; alkaline soil pink
MOISTURE: Most average situations
SEASONAL ASPECT: Flowering time in early summer
ZONE: 1, 2, 3
USES: Occasional flowering shrub, freestanding, interest, against tree in wooded area
CARE: Some pruning
PROPAGATION BY: Cuttings
NATIVE OF: e. United States
OTHER: Other Hydrangeas are:
H. arborescens; wild type; e. United States.
H. quercifolia; Oak-leaved Hydrangea; e. United States.
H. macrophylla; the Hortensia and Lace cup types; Japan.
H. paniculata var. grandiflora; Pee Gee Hydrangea; to 25′ tall; flower clusters change in color from white to greenish-purple; China; Japan.

Hypericum

Hypericum spp.

St. Johnswort Family

prolificum

H. lloydii

GROWTH HABIT: Evergreen and semievergreen shrubs
LEAVES: Opposite, relatively narrow, entire, sessile or nearly so, blunt or pointed at tips
FLOWERS: 5-parted, yellow, many stamens
FRUITS: Capsule
SEEDS: Tiny
LIGHT: Half to three-fourths sun
SOIL: Variable as to species
MOISTURE: As above
SEASONAL ASPECT: Summer flowers
ZONE: See below
USES: Ground cover, rockery, border, bed
CARE: None
PROPAGATION BY: Seedlings or stolons
NATIVE OF: See Below
OTHER: Ground cover or mat-forming species:

H. lloydii; s.e. coastal plain, sand hills; low, spreading shrub for dry, sandy, sunny spots; zone 2.

H. buckleyi; Carolina mountains; low spreading shrub for rockeries; zone 3.

H. calycinum; Europe and Asia Minor; leaves to 4″ long, purplish in winter; flowers to 3″ wide.

Shrubs from 2-6′ high:

H. prolificum; e. United States; a hardy mountain and piedmont species; very shrubby; narrow leaves; 3/4″ flowers; long blooming.

H. densiflorum; e. United States; similar to above but limited to the coastal plain.

H. frondosum; e. United States; somewhat smaller than above two but with larger flowers; zones 2, 3.

H. patulum and varieties; China; "Hidcote", "Henryi" and "Sungold" are popular varieties, each with fragrant flowers to 2″ in diameter; zones 1, 2, 3.

Idesia

Idesia polycarpa
Flacourtia Family

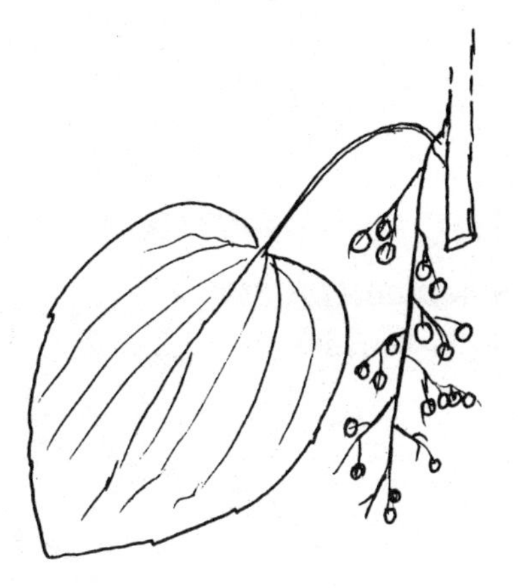

GROWTH HABIT: Small deciduous tree with whorls of horizontal branches
LEAVES: Heart-shaped, to 5″ long (rather like Catalpa), remotely toothed and long stalked
FLOWERS: Small, greenish-yellow, fragrant, in large panicles, no petals, male and female usually on separate trees
FRUITS: Orange-red berries
SEEDS: Few
LIGHT: Lots of sun
SOIL: Good productive type
MOISTURE: Dry
SEASONAL ASPECT: Colored berries in fall, if male tree is near
ZONE: 1, 2, 3
USES: Freestanding, shade, interest
CARE: None
PROPAGATION BY: Seeds
NATIVE OF: Taiwan, Japan, China

India Hawthorn

Rhaphiolepis indica

Rose Family

GROWTH HABIT: Evergreen shrub to 4′ high
LEAVES: Alternate, leathery, lanceolate to oblong, entire or toothed and to 3″ long
FLOWERS: Produced in profusions, pinkish
FRUITS: Bluish and about 1/4″ wide
SEEDS: 1-2
LIGHT: Full sun for best blossoms
SOIL: Fertile
MOISTURE: Moist
SEASONAL ASPECT: Late spring flowers
ZONE: 1, 2, 3
USES: Bed, hedge, foundation, occasional flowering shrub
CARE: None
PROPAGATION BY: Seedlings and cuttings
NATIVE OF: s. China

Indian Cherry, Buckthorn

Rhamnus spp.

Buckthorn Family

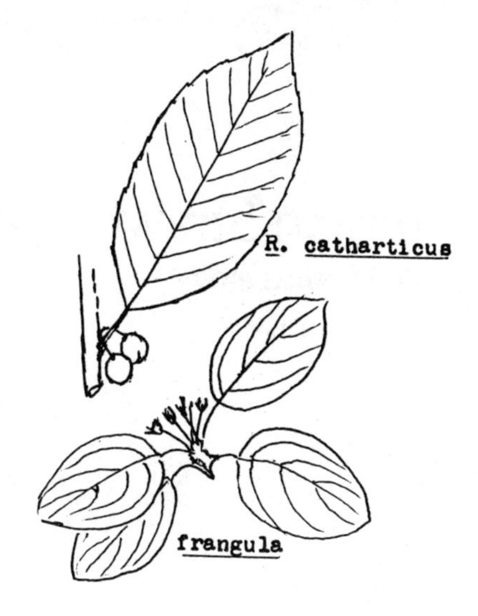

GROWTH HABIT: Deciduous shrubs or small trees

LEAVES: Alternate or opposite, elliptic to ovate, finely toothed to nearly entire, 1-5″ long, veins in pairs

FLOWERS: Small, greenish or yellowish, unisexual or bisexual, 4 or 5-parted

FRUITS: Globose, succulent, black at maturity and 1/4″ wide

SEEDS: 1

LIGHT: Partial or full sun

SOIL: Fertile loam

MOISTURE: Moist

SEASONAL ASPECT: None

ZONE: 1, 2, 3

USES: Occasional shrub, specimen

CARE: None

PROPAGATION BY: Seed

NATIVE OF: See below

OTHER: Indian Cherry, *R. caroliniana*; e. United States, leaves alternate with 8-10 pairs of veins, elliptic; flowers bisexual and 5-parted.

Buckthorn, *R. cathartica*; Europe, Africa, Asia; leaves opposite, with 3-5 pairs of veins, ovate; flowers unisexual and 4-parted.

Hedge Buckthorn, *R. frangula*, Europe, Asia; vigorous columnars type to 15′ high; its narrow habit makes it a good hedge plant, requiring only top clipping. *Rhamnus* bark has long been used as a laxative.

Indian Strawberry

Duchesnea indica

Rose Family

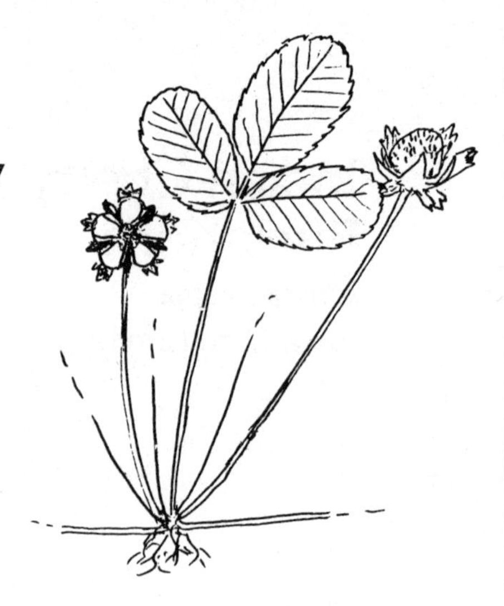

GROWTH HABIT: Evergreen herb spreading by runners to form mat
LEAVES: Compound with three toothed leaflets
FLOWERS: Yellow, 5 petals, solitary on long stalks, 1/2″ or more wide
FRUITS: Red, strawberry-like and to 1/2″ wide
SEEDS: Several
LIGHT: Partial shade
SOIL: Most any type
MOISTURE: Moist
SEASONAL ASPECT: Flowers and fruits throughout the summer
ZONE: 1, 2, 3
USES: Ground cover, to 6″ deep
CARE: None
PROPAGATION BY: Seeds and rooted stems
NATIVE OF: s. Asia

Indigo

Indigofera suffruticosa

Legume Family

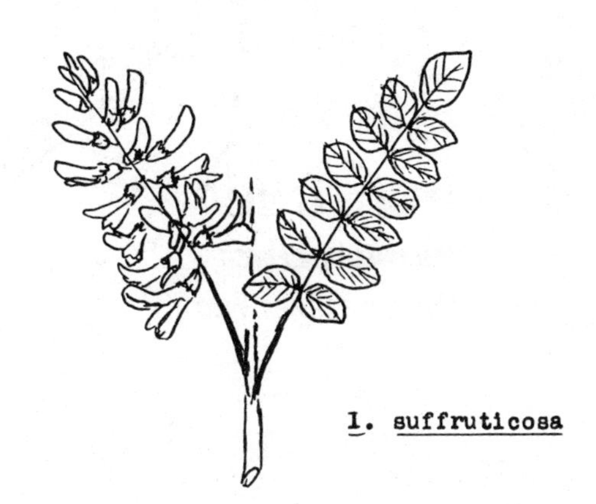

GROWTH HABIT: Sparingly branched, deciduous shrub to 5′ high
LEAVES: Alternate, pinnate, leaflets 9-17, thin, oblong and to about 1″ in length
FLOWERS: Very small, orange and in racemes
FRUITS: Stout curved pod 1/2″ long
SEEDS: Few
LIGHT: Full sun
SOIL: Fertile
MOISTURE: Moist
SEASONAL ASPECT: None
ZONE: 1
USES: Interest, bed
CARE: None
PROPAGATION BY: Seeds
NATIVE OF: w. Indies
OTHER: Once cultivated as the source of indigo.
I. tinctoria; with straight or almost straight seed pod. Also used as a source of indigo and found throughout the tropics.
I. kirilowii; China, Japan; deciduous shrub to 3′ high with racemes of 3/4″ long rose-colored flowers.

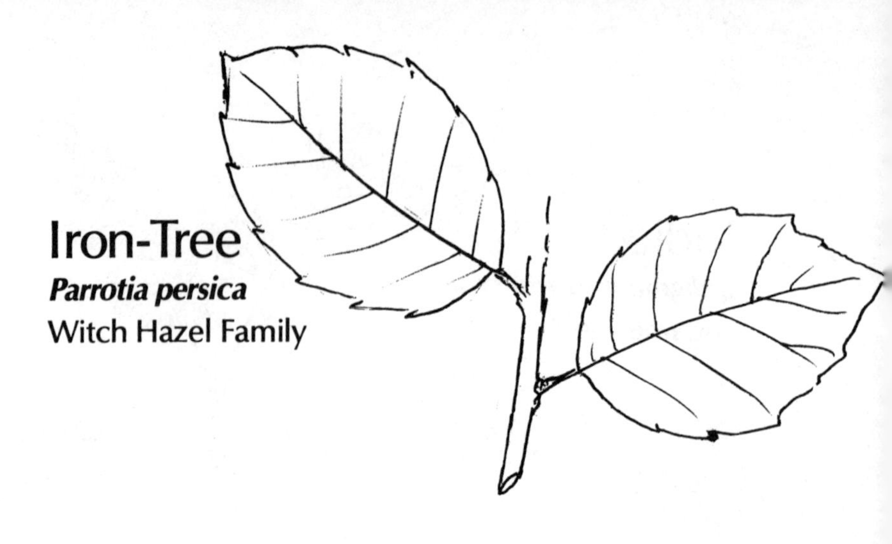

Iron-Tree

Parrotia persica

Witch Hazel Family

GROWTH HABIT: Large, deciduous shrub, stems mottled from peeling bark
LEAVES: Alternate, broadly ovate, sparsely toothed
FLOWERS: Clusters of purple-red stamens hanging like brushes from 1/4″ wide flower
FRUITS: Woody
SEEDS: Few and small
LIGHT: Lots of sun
SOIL: Any fertile, porous type
MOISTURE: Moist to dry
SEASONAL ASPECT: Very early spring flower and one of the finest shrubs or small trees for autumn colors
ZONE: (1), 2, 3
USES: Specimen
CARE: None
PROPAGATION BY: Cuttings
NATIVE OF: n. Persia, Caucasus
OTHER: A pendulous form exists

Japanese Bamboo

Polygonum cuspidatum

Smartweed Family

GROWTH HABIT: Stout little branched subshrub to 6′ high, forming large clumps by underground rhizomes; deciduous

LEAVES: Alternate, reddish when young, oval to ovate, to 5″ long and abruptly pointed

FLOWERS: Small, numerous in leaf axils and in terminal clusters, greenish-white

FRUITS: 3-sided achene in winged calyx

SEEDS: 1

LIGHT: Full or partial sun

SOIL: Most types

MOISTURE: Moist

SEASONAL ASPECT: None

ZONE: 1, 2 (3)

USES: Possibly as filler but is a pernicious spreader

CARE: None

PROPAGATION BY: Rhizomes

NATIVE OF: Japan

OTHER: Silver Lace Vine, *P. aubertii,* China; vigorous twiner with leaves to 2″ long and large clusters of fragrant whitish flowers.

Japanese Raisin-Tree

Hovenia dulcis

Buckthorn Family

GROWTH HABIT: Small, round-topped deciduous tree
LEAVES: Alternate, heart-shaped, long-pointed, and toothed
FLOWERS: Small, purplish, 5-parted and in axillary or terminal clusters
FRUITS: 1/4″ wide berry on an enlarged and edible stalk
SEEDS: 1-3
LIGHT: Full or lots of sun
SOIL: Fertile
MOISTURE: Dry to moist
SEASONAL ASPECT: None
ZONE: 1, 2, 3
USES: As small tree or may be kept as shrub by pruning
CARE: Little, if any
PROPAGATION BY: Scarified seeds or summer cuttings
NATIVE OF: Japan, China and the Himalayas
OTHER: Essentially a foliage plant

Winter Jasmine, Jessamine

Jasminum nudiflorum

Olive Family

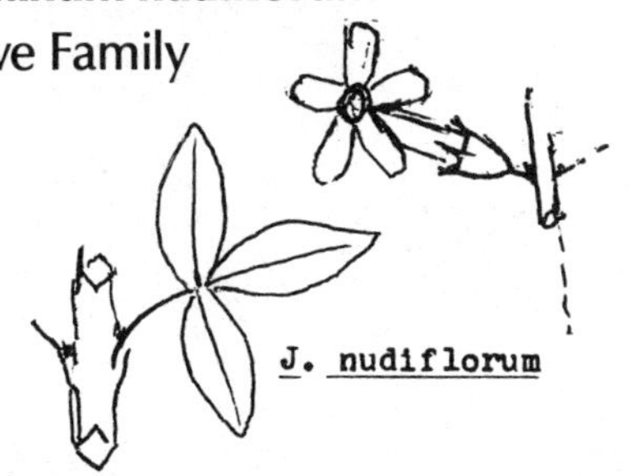

GROWTH HABIT: Deciduous shrub to 5′ with stiff 4-angled branches
LEAVES: Opposite, trifoliate, leaflets to 1″ long and entire
FLOWERS: Yellow, 4-parted and almost 1″ across
FRUITS: 2-lobed berry
SEEDS: 2
LIGHT: Full sun or nearly so
SOIL: Most any good productive type
MOISTURE: Moist to dry
SEASONAL ASPECT: Showy yellow flowers before the leaves
ZONE: 1, 2, 3
USES: Bed, foundation, unit arrangement
CARE: Some pruning
PROPAGATION BY: Cuttings
NATIVE OF: China
OTHER: Primrose Jasmine; *J. mesnyi*; w. China; evergreen, opposite leaves; slender arching branches and larger leaves and flowers than the Winter Jasmine.

Lady Jasmine; *J. humile*; Tropical Asia; evergreen alternate leaves; very slender arching stems; finer textured foliage and smaller flowers than either of the above.

Poets Jessamine, *J. officinale*; deciduous shrub with long weak stems that require support; opposite, pinnate leaves with 5-7 leaflets, the terminal largest; flowers white, fragrant and to 1″ wide.

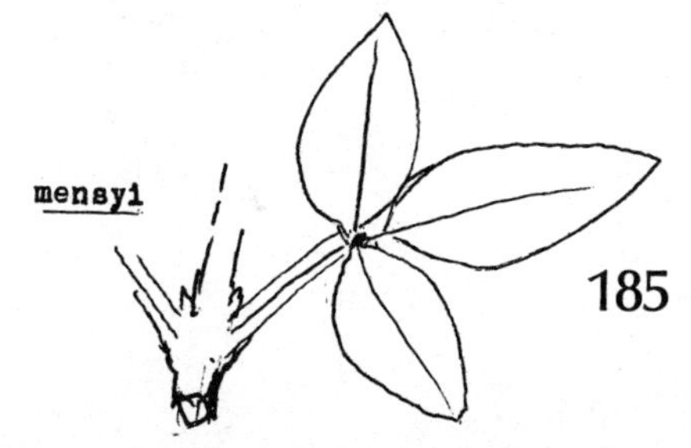

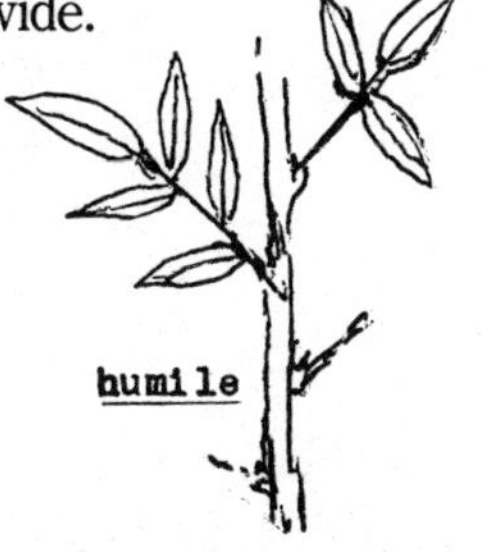

Jerusalem-Cherry

Solanum pseudo-capsicum

Nightshade Family

GROWTH HABIT: Semievergreen branching shrub to 3′ high
LEAVES: Alternate, oblong to oblanceolate, to 3″ long, glabrous and entire
FLOWERS: White, 1/2″ wide and axillary
FRUITS: Cherry-size, red or yellow and long-persisting
SEEDS: Numerous
LIGHT: Full or partial sun
SOIL: Any productive type
MOISTURE: Moist
SEASONAL ASPECT: Late summer and fall fruits
ZONE: 1, 2
USES: Patio, terrace, border, pot
CARE: Only moderately cold-hardy
PROPAGATION BY: Seeds and seedlings
NATIVE OF: Mediterranean region (?)
OTHER: Fruit very poisonous.

Jerusalem-Thorn

Parkinsonia aculeata

Legume Family

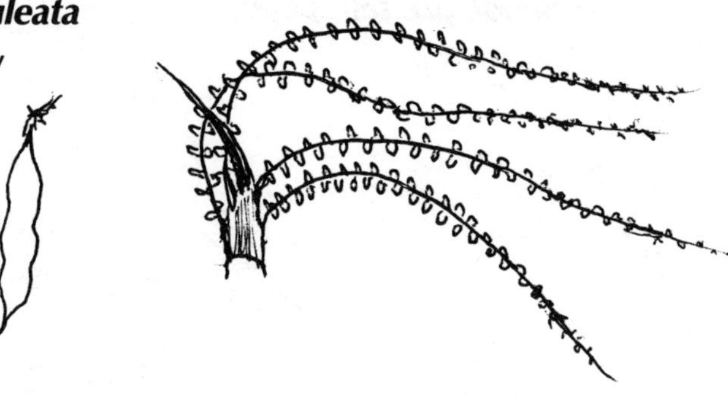

GROWTH HABIT: Small, deciduous tree, usually with slender, drooping zigzag branches
LEAVES: Alternate, bipinnate, very short petiole, very short spine-tipped rachis, 1-2 pairs of twig-like pinnae bearing numerous very small leaflets
FLOWERS: Yellow and on drooping racemes 2-5″ long
FRUITS: Narrow pod to 5″ long and constricted between seeds
SEEDS: Several
LIGHT: Full sun
SOIL: Sandy loam
MOISTURE: Moist
SEASONAL ASPECT: None
ZONE: 1
USES: Interest, background
CARE: None
PROPAGATION BY: Seeds
NATIVE OF: Tropical America
OTHER: Suitable only for coastal area from South Carolina south, because of climate.

Jetbead

Rhodotypos tetrapetala

Rose Family

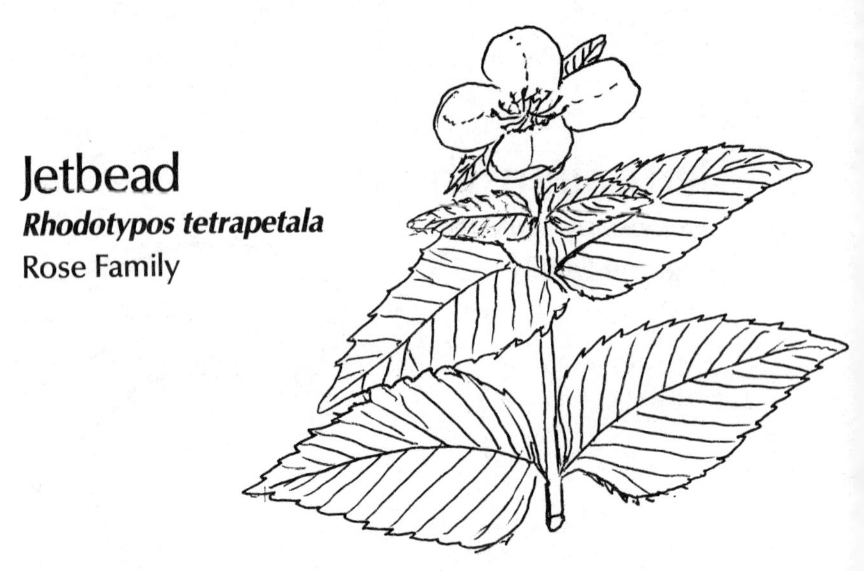

GROWTH HABIT: Deciduous shrub to 5′ high with reddish stems
LEAVES: Opposite, ovate, pointed, doubly sharp-toothed, rough above, to 4″ long and veiny
FLOWERS: White, 1″ or more across and on shoots of the season
FRUITS: Shining black, 1/4″ wide and long, persistent as are the enlarged spreading calyx lobes
SEEDS: 1 and rather large
LIGHT: Full or lots of sun
SOIL: Fertile sandy soil
MOISTURE: Moist
SEASONAL ASPECT: Late spring flowers and winter fruits
ZONE: 1, 2, (3)
USES: Interest, specimen, unit arrangement
CARE: None
PROPAGATION BY: Seedlings and cuttings
NATIVE OF: Japan

Jointweed

Polygonella americana

Smartweed Family

GROWTH HABIT: Deciduous subshrub to 3′ high
LEAVES: Linear and less than 1″ long
FLOWERS: White, occasionally pink, small, numerous and forming large branched arrangement
FRUITS: 3-angled achene
SEEDS: 1
LIGHT: Full sun
SOIL: Light sandy
MOISTURE: Dry
SEASONAL ASPECT: Fall flowers
ZONE: 1, 2
USES: Bank, border
CARE: None
PROPAGATION BY: Seeds
NATIVE OF: s.e. United States
OTHER: Gaining popularity

Juniper

Juniperus spp.

Cypress Family

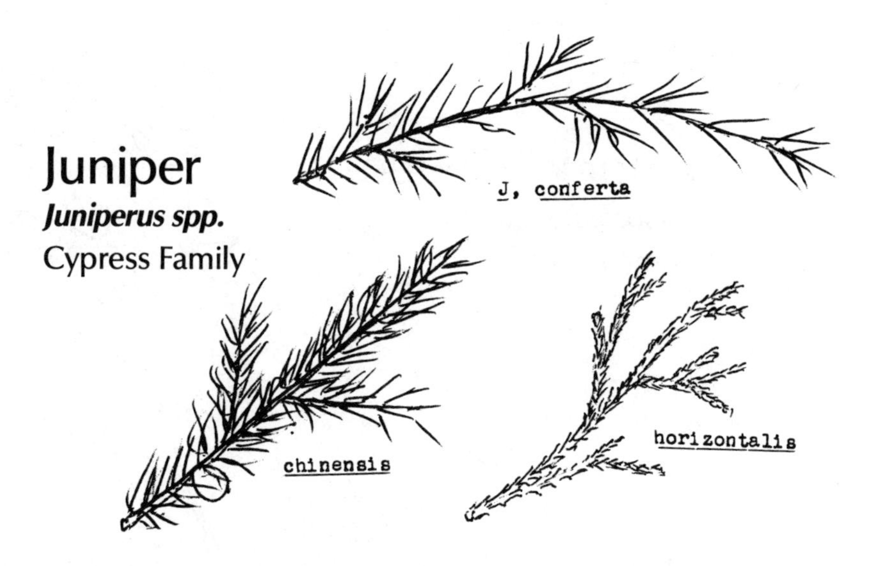

GROWTH HABIT: A number of species of evergreen trees and shrubs but available mostly as shrubs which vary greatly in size, form and color.

LEAVES: Mostly opposite and needle-like or scale-like, of varying lengths and colors which range from intense blue-white through shades of green to yellow.

FLOWERS: Minute and inconspicuous

FRUITS: Bluish berries, when produced

SEEDS: 1-6 per fruit

LIGHT: Full or partial sun

SOIL: Loams with average or better fertility

MOISTURE: Moist

SEASONAL ASPECT: None

ZONE: 1, 2, 3

USES: Border, planter, terrace, bank, edging, unit arrangement and winter appeal

CARE: Some pruning and watering

PROPAGATION BY: Cuttings and seedlings

NATIVE OF: See below

OTHER: *J. conferta,* Shore Juniper, Japan; prostrate with prickly leaves.

J. communis, Common Juniper, Europe, Asia and North America. Also native variety *depressa.*

J. chinensis, Chinese Juniper, China, very many clones and habit-forms are cultured.

J. horizontalis, Creeping Juniper, n. United States and s. Canada. The above and others are available in many forms.

Kentucky Coffee-Tree

Gymnocladus dioica

Legume Family

GROWTH HABIT: Rather small, deciduous tree
LEAVES: Alternate, bipinnate, to 3′ long with 3-7 pairs of main divisions (pinnae) each with several pairs of ovate leaflets
FLOWERS: Greenish white in panicles
FRUITS: Hard, dark-colored pods, 3-6″ long and flattened
SEEDS: Few and large
LIGHT: Full or mostly full sun
SOIL: Fertile
MOISTURE: Moist
SEASONAL ASPECT: None
ZONE: 2, 3
USES: Interest, background, freestanding
CARE: None
PROPAGATION BY: Seeds
NATIVE OF: Ontario to Wisconsin, south to Alabama and Oklahoma

Kerria

Kerria japonica

Rose Family

GROWTH HABIT: Arching, little-branched, deciduous shrub, 4-8′ high with green branches

LEAVES: Alternate, to 3″ long, toothed, broadly lanceolate, and ong pointed

FLOWERS: 1 1/2″ wide, golden-yellow and usually with very many petals

FRUITS: 3-8 achenes per flower

SEEDS: 1 per achene

LIGHT: Full or partial sun

SOIL: Fertile

MOISTURE: Moist

SEASONAL ASPECT: Flowering late spring to autumn

ZONE: 1, 2, 3

USES: Occasional flowering shrub

CARE: Tends to be a colony-former

PROPAGATION BY: Colony division

NATIVE OF: Japan

OTHER: Several varieties cultivated.

Kiwi-Berry

Actinidia chinensis

Actinidia Family

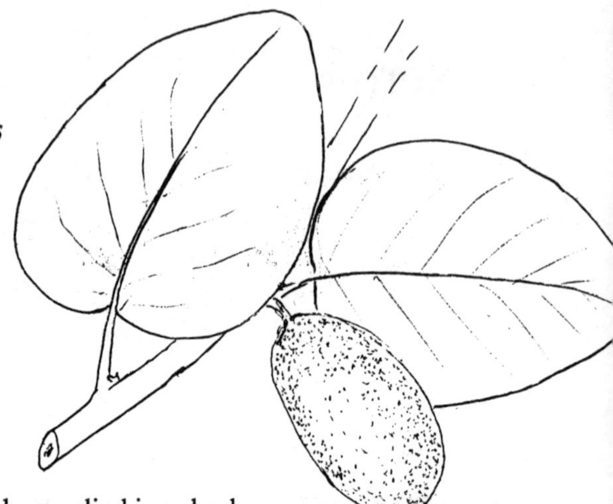

GROWTH HABIT: Deciduous climbing shrub
LEAVES: Alternate, broadly heart-shaped and to 6″ long
FLOWERS: Male and female on separate plants, white, turning to buff with age, 1 1/2″ wide
FRUITS: Somewhat fuzzy berry 2-3″ long—edible
SEEDS: Many and very small
LIGHT: Full sun
SOIL: Fertile loam
MOISTURE: Moist
SEASONAL ASPECT: None
ZONE: 1, 2
USES: Shade, fruit, fence
CARE: Pruning
PROPAGATION BY: Cuttings
NATIVE OF: China
OTHER: Once thought to have been from Australia.

Ladies'-Eardrops

Brunnichia cirrhosa

Smartweed Family

GROWTH HABIT: Deciduous, partly woody vine, high climbing by tendrils
LEAVES: Alternate, ovate and to 4″ long
FLOWERS: Several, greenish, 1″ long and terminal, or nearly so
FRUITS: Nutlets, 1/4″ long and 3-angled
SEEDS: 1
LIGHT: Full or partial sun
SOL: Rich alluvium
MOISTURE: Moist
SEASONAL ASPECT: None
ZONE: 1
USES: Specimen, interest, arbor
CARE: None
PROPAGATION BY: Seeds
NATIVE OF: s.e. United States coastal plain

Lantana

Lantana camara

Vervain Family

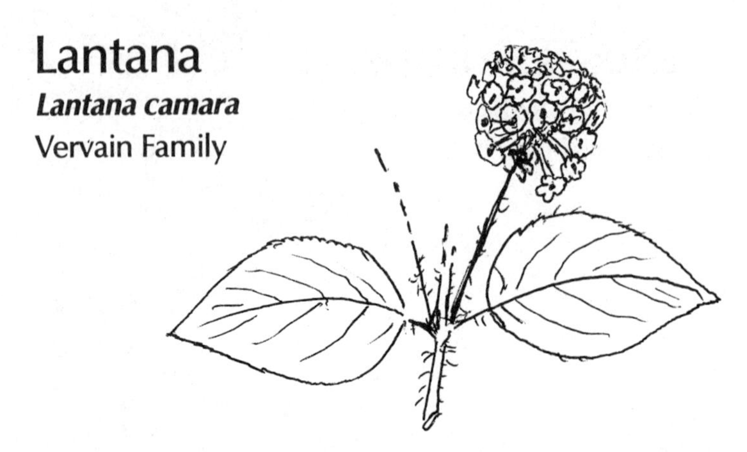

GROWTH HABIT: Coarse-hairy, deciduous shrub to 5′ with weak recurved prickles on the stem angles

LEAVES: Opposite, ovate, coarsely pubescent and shallowly toothed

FLOWERS: Small but numerous in head-like clusters, pink, orange, yellow or lavender, varying with age

FRUITS: Blue or black berries

SEEDS: 2

LIGHT: Full or partial sun

SOIL: Most any type

MOISTURE: Moist

SEASONAL ASPECT: Flowers from late spring into fall

ZONE: 1, 2 (3)

USES: Interest, blender

CARE: None

PROPAGATION BY: Seeds and seedlings

NATIVE OF: s. Georgia to central America

OTHER: One or more varieties or species, some without prickles, are cultivated and occasionally escaped. Fruits toxic.

Lavender

Lavandula officinalis

Mint Family

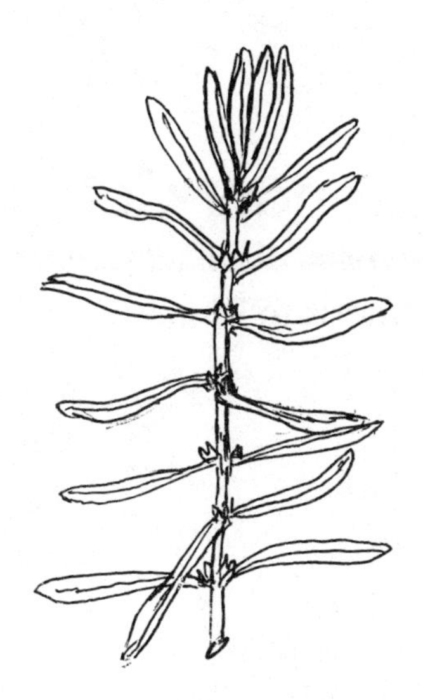

GROWTH HABIT: Low compact, grayish, aromatic, evergreen shrub to 2′ high
LEAVES: Opposite, or appearing clustered, linear to lanceolate, entire, white-hairy and to 1 1/2″ long, about 8X as long as wide
FLOWERS: Lavender, pink or white and in terminal whorls
FRUITS: 4 nutlets per flower
SEEDS: 1 seed per nutlet
LIGHT: Full or lots of sun
SOIL: Fertile and porous
MOISTURE: Dry to moist
SEASONAL ASPECT: Mid-summer flowers and their delicate fragrance
ZONE: 1, 2, 3
USES: Evergreen border, rockery
CARE: None
PROPAGATION BY: Seedlings
NATIVE OF: Mediterranean region
OTHER: The flowers remain fragrant long after drying and were used to perfume bath water.

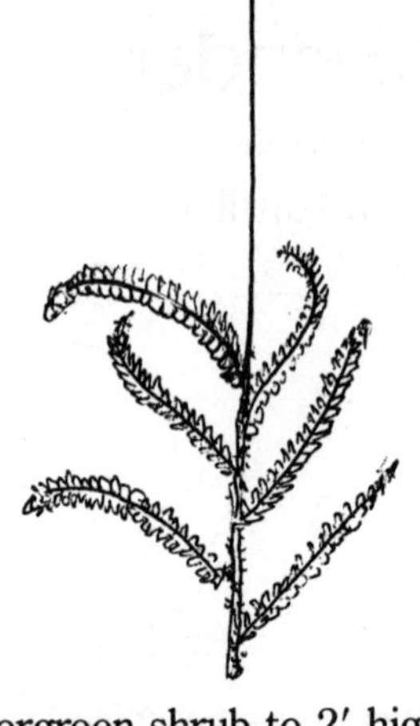

Lavender-Cotton

Santolina chamaecyparissus

Composite Family

GROWTH HABIT: Much branched evergreen shrub to 2′ high, aromatic and silvery-gray hairy
LEAVES: Alternate, about 1′ long and pinnately finely lobed
FLOWERS: Tiny yellow and in globular heads 1/2″ wide
FRUITS: Achene
SEEDS: 1
LIGHT: Full sun
SOIL: Tolerant
MOISTURE: Dry
SEASONAL ASPECT: None
ZONE: 1, 2, 3
USES: Edging, border, bank
CARE: None
PROPAGATION BY: Seedlings
NATIVE OF: Mediterranean region

Leather Leaf

Cassandra calyculata (Chamaedaphne)

Heath Family

GROWTH HABIT: Much branched evergreen shrub to 4′ high
LEAVES: Alternate, elliptic, to 2″ long and rusty beneath
FLOWERS: White, urn-shaped, 1/4″ long and in terminal leafy racemes
FRUITS: Woody capsule
SEEDS: Very small
LIGHT: Partial shade
SOIL: Acid type
MOISTURE: Moist to wet
SEASONAL ASPECT: Spring flowers
ZONE: 1, 2
USES: Evergreen shrub for wet sites
CARE: No special
PROPAGATION BY: Seedlings and roots
NATIVE OF: s.e. United States, n. America, n. Europe, n. Asia

Leatherwood

Dirca palustris

Mesereum Family

GROWTH HABIT: Freely branched, deciduous shrub with jointed twigs
LEAVES: Alternate, entire, mostly obovate and short petioled
FLOWERS: Barely 1/2″ long, pale yellow, in small groups along the twigs before the leaves
FRUITS: 1/4″ long red drupe
SEEDS: 1
LIGHT: Mostly shade
SOIL: Rich loam
MOISTURE: Moist or wet
SEASONAL ASPECT: None
ZONE: 2, 3
USES: Specimen
CARE: Little
PROPAGATION BY: Seedlings
NATIVE OF: e. United States

Leptodermis

Leptodermis oblonga

Madder Family

GROWTH HABIT: Deciduous shrub to 3′ high
LEAVES: Oval, rough above and to 1″ long
FLOWERS: Lilac-purple, 1/2″ long and Lilac-like
FRUITS: Capsule
SEEDS: Several
LIGHT: Full or partial sun
SOIL: Any good type
MOISTURE: Moist
SEASONAL ASPECT: Late summer flowers
ZONE: 1, 2, 3
USES: Specimen
CARE: Occasional trimming
PROPAGATION BY: Cuttings
NATIVE OF: n. China
OTHER: An excellent bloomer

Lespedeza

Lespedeza bicolor

Legume Family

GROWTH HABIT: Deciduous shrub to 8′ high
LEAVES: Alternate, 3-foliate, and, to 1 1/2″ long with rounded tips
FLOWERS: Rose-purple and white, 1/2″ long and numerous
FRUITS: 1-seeded, pubescent pod about 1/4″ long
SEEDS: 1
LIGHT: Full or partial sun
SOIL: Tolerant
MOISTURE: Moist
SEASONAL ASPECT: Mid- and late-summer flowers
ZONE: 1, 2, 3
USES: Margin, border
CARE: None
PROPAGATION BY: Seedlings
NATIVE OF: Japan and n. China
OTHER: Much planted for ground-feeding birds. *L. cyrtobotrya*; Japan, Korea; similar to above but a profuse bloomer in late summer.

Common Lilac

Syringa vulgaris

Olive Family

GROWTH HABIT: Deciduous shrub spreading by roots

LEAVES: Opposite, ovate, long pointed, entire 2-4″ long and with a whitened appearance beneath

FLOWERS: Lilac or white, each a little over 1/4″ long, fragrant and produced usually in paired panicles to 8″ long

FRUITS: Small leathery capsule

SEEDS: Several, winged

LIGHT: Partial sun

SOIL: Most any fertile type

MOISTURE: Moist to dry

SEASONAL ASPECT: Early spring flowers, following Dogwood

ZONE: (2), 3

USES: Occasional flowering shrub, border

CARE: Yearly pruning and periodic renewal pruning that removes half or all of the top 6″ or so above ground

PROPAGATION BY: Cuttings

NATIVE OF: e. Europe

OTHER: Among the 500 or so species and varieties of Lilacs, the most popular and common by far are the more than 400 named varieties of *S. vulgaris*. Hot summers in the south limit widespread use.

S. laciniata; n.w. China; particularly interesting because its leaves are deeply pinnately lobed.

Liriope

Liriope spp.

Lily Family

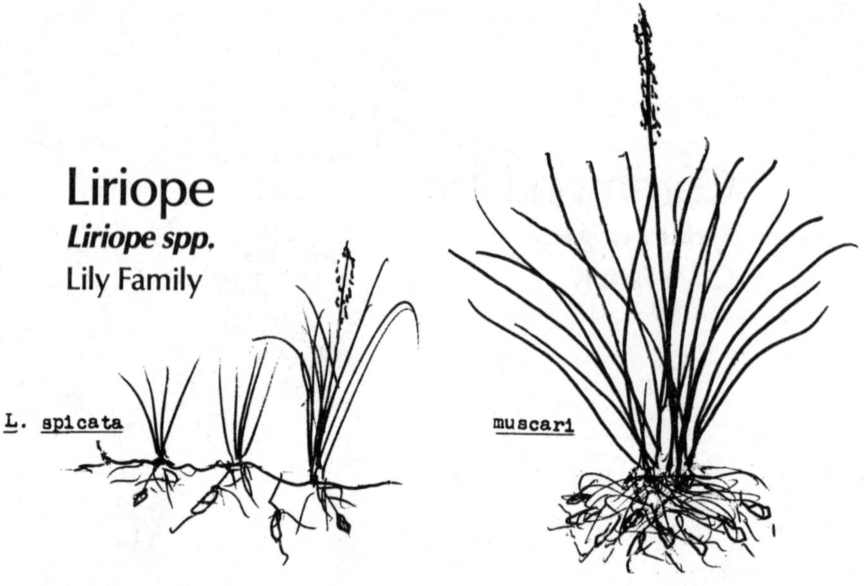

GROWTH HABIT: Grass-leaved evergreens 6-12″ high, frequently tuber-forming

LEAVES: 1/4 to 3/4″ wide, arising from ground level or below

FLOWERS: Small, white to lilac, borne in racemes or panicles on leafless stem half as high or equaling leaves; flowers short pediceled

FRUITS: Blue and about 1/4″ wide

SEEDS: 1-3

LIGHT: Light or shade

SOIL: Tolerant

MOISTURE: Moist

SEASONAL ASPECT: None

ZONE: 1, 2, 3

USES: Border, terrace, edging, sidewalk, pot

CARE: None

PROPAGATION BY: Roots

NATIVE OF: China, Japan

OTHER: Monkey-Grass, *L. spicata*; narrow leaves, 1/4″ wide, pale flowers, mat-former; spreads vigorously by roots and seeds.

Border-Grass, *L. muscari*; leaves 1/2-3/4″ wide, lilac flowers, clump-former; one variety has striped leaves.

Black Locust

Robinia pseudoacacia

Legume Family

GROWTH HABIT: Medium-sized, deciduous tree with prickly branches due to presence of stipular spines; bark deeply furrowed
LEAVES: Alternate, pinnate, 7-17 elliptic leaflets to 1″ long
FLOWERS: 3/4″ long, white, fragrant, on drooping racemes 3-6″ long
FRUITS: Flat, thin pod, 2-4″ long, remaining on tree into winter
SEEDS: Several
LIGHT: Partial or full sun
SOIL: Fertile porous loam
MOISTURE: Moist
SEASONAL ASPECT: White flowers soon after leaves appear
ZONE: 1, 2, 3
USES: Streets, sidewalk, freestanding
CARE: None
PROPAGATION BY: Seedlings or cuttings
NATIVE OF: e. North America
OTHER: The wood is very hard and extremely rot-resistant; Several horticultural varieties exist.

Honey Locust

Gleditsia triacanthos

Legume Family

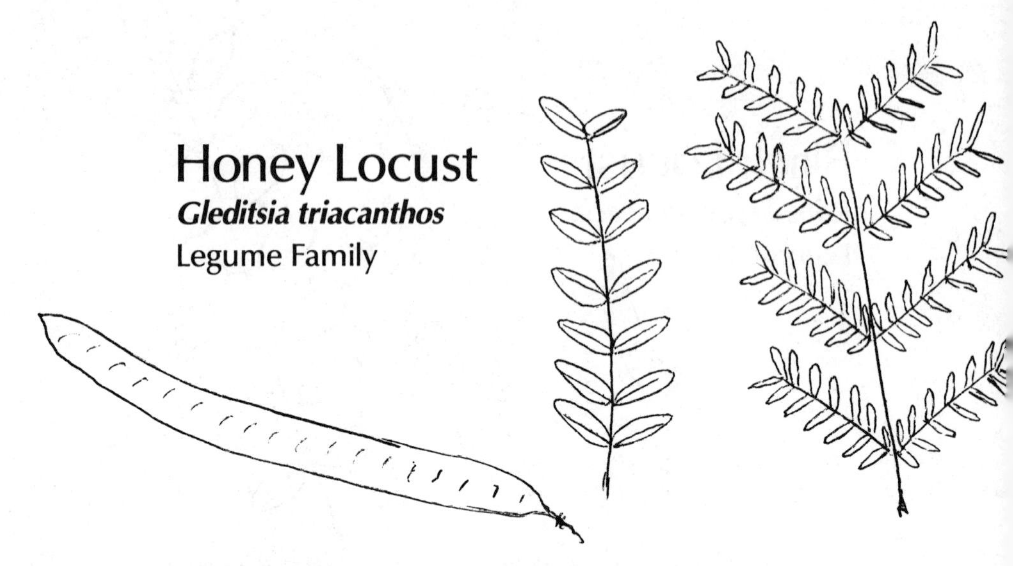

GROWTH HABIT: Large, deciduous tree with simple or branched thorns to 5″ long
LEAVES: Alternate, pinnate and often also bipinnate, leaflets many, even-numbered and 1/2″ long
FLOWERS: White, frequently unisexual on any particular raceme
FRUITS: Dark brown, twisted pods, 10″ or more long, pulpy on one side and hard and woody on the other
SEEDS: Several
LIGHT: Full or mostly full sun
SOIL: Tolerant
MOISTURE: Moist
SEASONAL ASPECT: None
ZONE: 1, 2, 3
USES: Background tree for thin shade
CARE: None
PROPAGATION BY: Seeds
NATIVE OF: e. United States
OTHER: A thornless variety is cultivated as the Moraine Locust. The pulpy part of the pod is somewhat edible and with Persimmons was the source of Locust-Persimmon beer.

Loquat, Japan Plum

Eriobotrya japonica

Rose Family

GROWTH HABIT: Large evergreen shrub or small tree
LEAVES: Alternate, large, stiff, obovate, toothed, dark green above, densely wooly beneath
FLOWERS: Whitish, 1/2″ wide, in short, wooly panicles late in the year
FRUITS: 1 1/2″ long, yellow, ripening the next spring; edible
SEEDS: Several
LIGHT: Full or partial sun
SOIL: Tolerant
MOISTURE: Moist
SEASONAL ASPECT: None
ZONE: 1, 2
USES: Shrub border, unit arrangement, background fruit
CARE: None, except possible pruning
PROPAGATION BY: Seedlings
NATIVE OF: Japan

Loropetalum

Loropetalum chinense

Witch Hazel Family

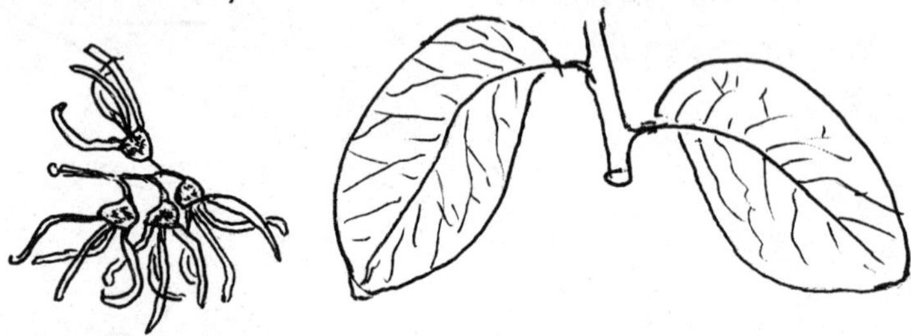

GROWTH HABIT: Much branched, evergreen shrub to 8′, twigs rusty pubescent

LEAVES: Alternate, ovate, short petioled, entire and somewhat unequal sided at base

FLOWERS: White to greenish, several together at ends of short branches; petals 4, long and strap-shaped

FRUITS: Woody capsule

SEEDS: 2

LIGHT: Sun or partial shade

SOIL: Fertile

MOISTURE: Moist

SEASONAL ASPECT: Early spring flowers

ZONE: 1, 2, 3

USES: Flowering shrub, foundation, border

CARE: Some pruning may be desired

PROPAGATION BY: Seedlings and cuttings

NATIVE OF: China and India

Maackia

Maackia amurensis

Legume Family

GROWTH HABIT: Small deciduous tree rather similar to Yellowwood
LEAVES: Alternate, pinnate with 7-11 leaflets, each 2-3″ long and toothless
FLOWERS: White, 1/2″ long and in elongated clusters
FRUITS: Short, flat pods
SEEDS: Few
LIGHT: Lots of sun
SOIL: Moderately fertile
MOISTURE: Dry
SEASONAL ASPECT: None
ZONE: 2, 3
USES: Freestanding, shade, specimen
CARE: None
PROPAGATION BY: Seeds, seedlings
NATIVE OF: Manchuria

Great Magnolia, Bull Bay

Magnolia grandiflora

Magnolia Family

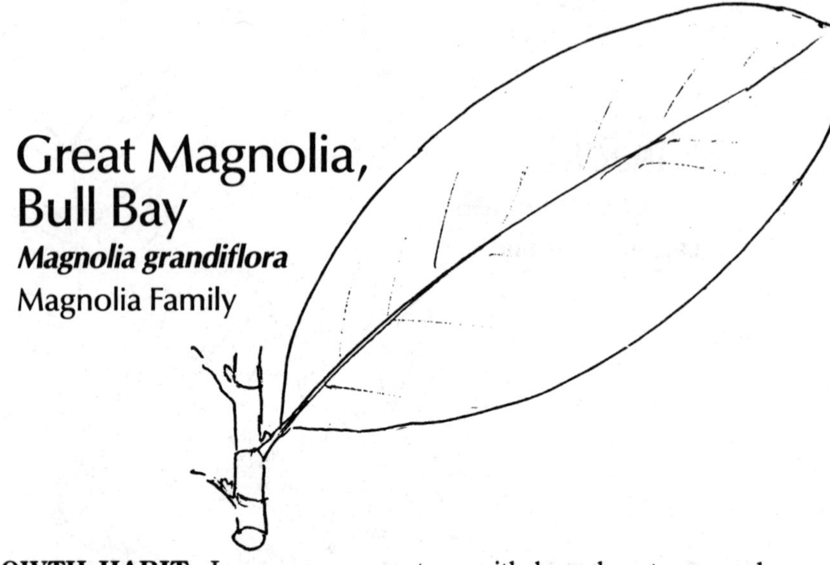

GROWTH HABIT: Large evergreen tree with branches to ground level and forming very dark shade
LEAVES: Alternate, thick, with or without purple pubescence beneath, entire, shining green above, 6-8″ long
FLOWERS: White and 5″ or more across
FRUITS: A cone-like structure
SEEDS: Red and 1 per follicle
LIGHT: Full sun or partial shade
SOIL: Fertile
MOISTURE: Moist
SEASONAL ASPECT: Flowering time in summer
ZONE: 1, 2, 3
USES: Freestanding, windbreak
CARE: None
PROPAGATION BY: Seedlings
NATIVE OF: s.e. United States
OTHER: A great favorite

Jujube

Ziziphus jujuba

Buckthorn Family

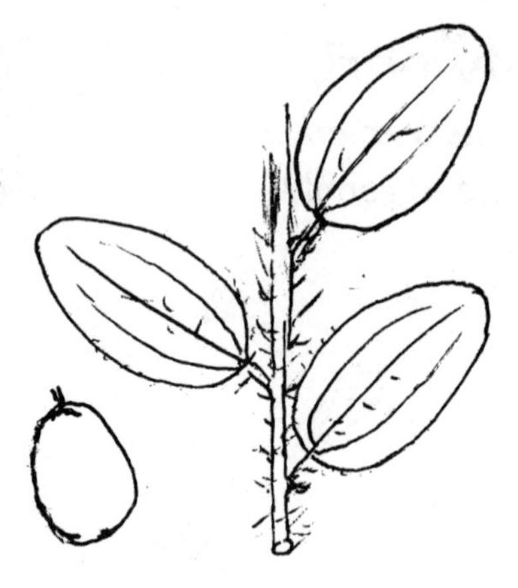

GROWTH HABIT: Deciduous shrub to 8′ or more high; spiny and spineless varieties
LEAVES: Alternate, ovate, to 2″ long and toothed
FLOWERS: Small, yellow, axillary and 5-parted
FRUITS: Brown or red drupe about 1″ long; edible
SEEDS: 1
LIGHT: Full sun
SOIL: Fertile loam
MOISTURE: Dry to moist
SEASONAL ASPECT: None
ZONE: 1, 2
USES: Interest, specimen
CARE: Some pruning
PROPAGATION BY: Seeds
NATIVE OF: e. Europe, s. and e. Asia

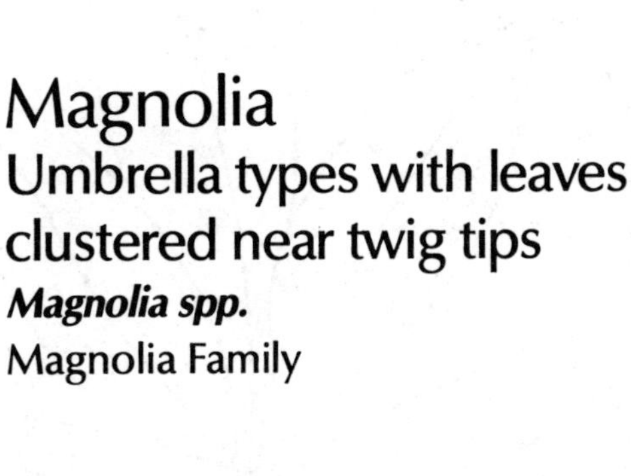

Magnolia
Umbrella types with leaves clustered near twig tips
Magnolia spp.
Magnolia Family

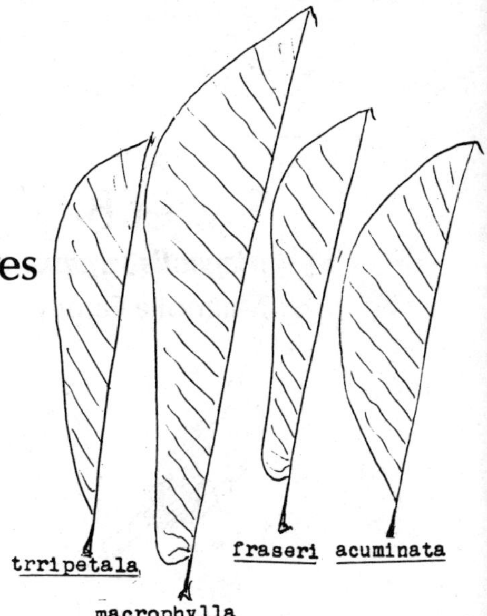

GROWTH HABIT: Medium to large deciduous trees
LEAVES: Large, 8″ - 3′ long, thin, crowded toward twig tips
FLOWERS: Large, yellowish or white, often high
FRUITS: Cone-like and 2-4″ long
SEEDS: Red
LIGHT: Partial or full sun
SOIL: Fertile loam
MOISTURE: Moist
SEASONAL ASPECT: Flowering time in late spring
ZONE: 1, 2, 3
USES: wherever deciduous trees are desired
CARE: None
PROPAGATION BY: Seedlings
NATIVE OF: s.e. United States
OTHER: Umbrella Tree, *M. tripetala*; leaves 8-15″ long; flowers greenish-yellow and slightly malodorus; petals 2-3″ long and spreading.

Large-leaved Umbrella-Tree, *M. macrophylla*; leaves 15-36″ long; flowers white, fragrant; petals to 8″ long and spreading. Our largest flower.

Ear-leaved Umbrella Tree, *M. fraseri*; leaves crowded at twig tips, to 15″ long, obovate and with ear-lobes at bases; flowers white, fragrant and 2-3″ wide; Zone 3.

Cucumber Tree, *M. acuminata*; large tree; leaves broadly elliptic and to 8″ long; flowers greenish-yellow, 2-3″ wide and not showy; unique fruit somewhat cucumber-like; Zones 2, 3.

Sweet Bay

Magnolia virginiana

Magnolia Family

GROWTH HABIT: Semievergreen shrub or slender tree

LEAVES: 3-5″ long, elliptic, white beneath and fragrant when crushed, scattered on twigs

FLOWERS: White, petals 2-3″ long, expanding widely only in age, fragrant

FRUITS: Cone-like structure 1 1/2″ long

SEEDS: Bright red

LIGHT: Partial shade

SOIL: Silt or silty loam

MOISTURE: Moist to wet

SEASONAL ASPECT: Late spring flowers

ZONE: 1, 2

USES: Low, wet border, interest

CARE: None

PROPAGATION BY: Seedlings

NATIVE OF: s.e. United States

OTHER: Sometimes grown in surprisingly dry places but where water is amply and regularly supplied during establishment.

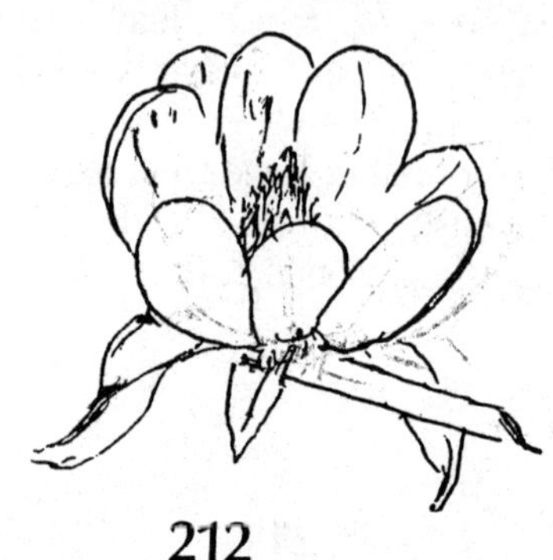

Magnolia
(Introduced)
Magnolia spp.
Magnolia Family

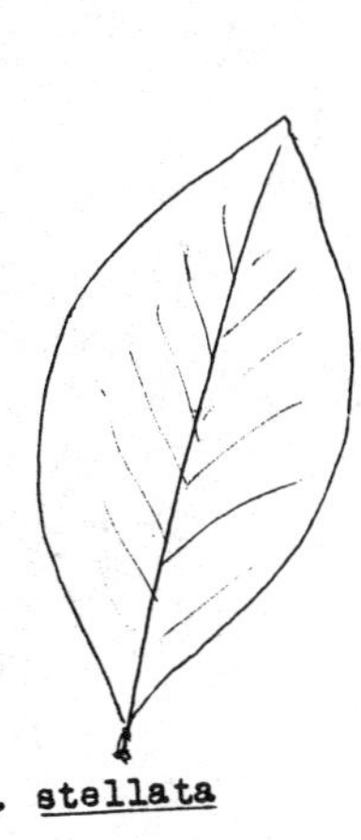

GROWTH HABIT: Deciduous shrubs
LEAVES: Alternate, scattered on twigs
FLOWERS: Showy, before the leaves, purple, rose or white
FRUITS: Interrupted cone-like strucutres, 1-3″ long
SEEDS: Scarlet
LIGHT: Partial shade or sun
SOIL: Fertile
MOISTURE: Moist
SEASONAL ASPECT: Early spring flowering time
ZONE: 1, 2
USES: Occasional shrub
CARE: None
PROPAGATION BY: Seedlings or cuttings
NATIVE OF: Japan
OTHER: Japanese Magnolia, *M. soulangeana*; large shrub; flowers purplish-rose to white, bell-shaped and 3-5″ wide; many varieties, all prolific bloomers.

Starry Magnolia, *M. stellata*; Japan; smaller than above; flowers white; petals narrow and about 12 in number, widely spreading and each about 1 1/2″ long.

M. liliflora; deciduous tree-like shrub to 10′ high; leaves obovate to oval and to 6″ long; flowers lily-like, purplish outside, white within and to 3″ long; before the leaves.

Male-Berry

Lyonia ligustrina

Heath Family

GROWTH HABIT: Deciduous shrub to 6′ high
LEAVES: Alternate, obovate to elliptic, to 2″ long and minutely toothed
FLOWERS: White, globose, less than 1/4′ long and in elongate clusters
FRUITS: Very small capsules
SEEDS: Tiny
LIGHT: Partial shade
SOIL: Most types
MOISTURE: Moist or wet
SEASONAL ASPECT: Summer flowers
ZONE: 1, 2, 3
USES: Specimen or informal shrub border
CARE: None
PROPAGATION BY: Root sprouts
NATIVE OF: e. United States
OTHER: Somewhat colony forming by root sprouts.

Manettia

Manettia glabra

Madder Family

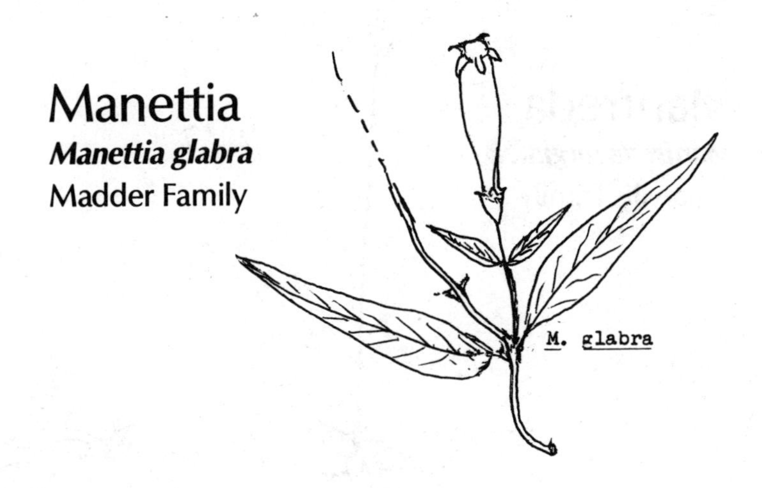

GROWTH HABIT: Slender evergreen vine to 5′ high
LEAVES: Opposite, lanceolate to ovate, entire and glabrous
FLOWERS: 1 1/2″ long, trumpet-shaped and red, corolla lobes small
FRUITS: Small capsule
SEEDS: Many
LIGHT: Full or partial sun
SOIL: Sandy
MOISTURE: Moist
SEASONAL ASPECT: Summer flowers
ZONE: 1, 2
USES: Wall, fence, post, pot
CARE: To be trained, as it is a weak climber
PROPAGATION BY: Cuttings
NATIVE OF: s. America
OTHER: *M. bicolor*; Brazil; similar but with broader leaves and flowers yellow within.

Manfreda

Manfreda virginica

Amaryllis Family

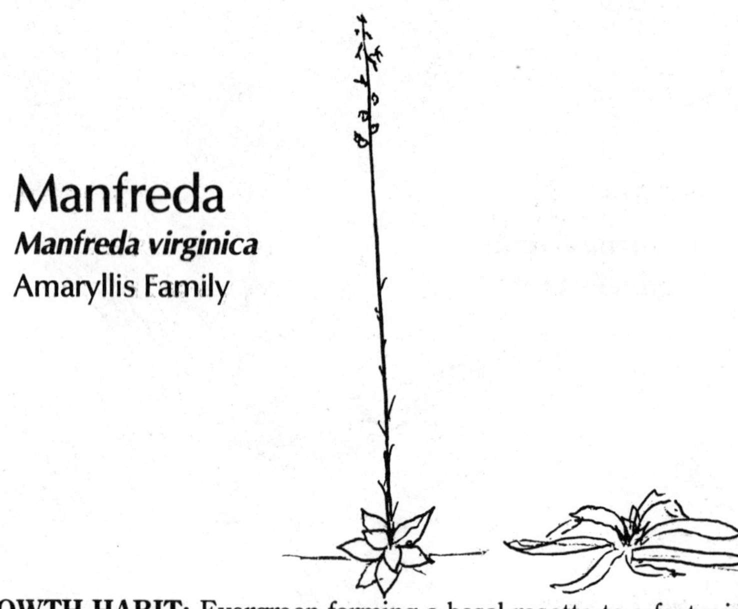

GROWTH HABIT: Evergreen forming a basal rosette to a foot wide and with occasional root sprouts

LEAVES: Succulent, broadly to narrowly lanceolate, sometimes purple or light green spotted

FLOWERS: Small yellowish, drooping and raised in very slender stalks to 5′

FRUITS: 1/2″ wide globose capsules

SEEDS: Several and flattened

LIGHT: Partial or full sun

SOIL: Most any type

MOISTURE: Moist to dry

SEASONAL ASPECT: Summer flower stalks unusual for their slender height

ZONE: 1, 2

USES: Specimen

CARE: None

PROPAGATION BY: Seeds and root sprouts

NATIVE OF: s.e. United States

OTHER: Two types exist, one of wet habitats and with dark green linear leaves to 1′ long; the other of dry habitats and with mottled lanceolate leaves to 6″ long.

Maple

Acer spp.

Maple Family

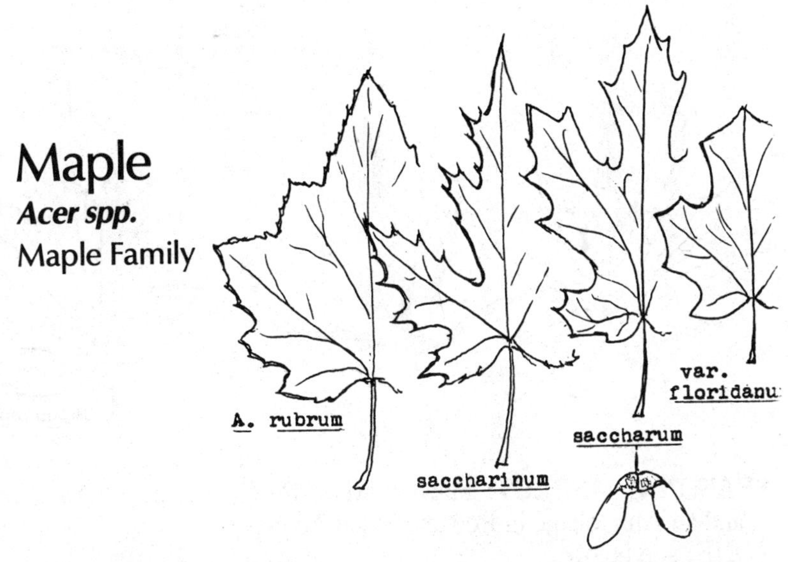

GROWTH HABIT: Deciduous trees and shrubs, some with male and female flowers on different plants. The horticultural importance of this genus is second only to Oak.

LEAVES: Opposite, petioled, palmately 3-7 or more lobed

FLOWERS: Small, clustered, in some species before the leaves and then, usually red

FRUITS: Nutlets borne in pairs, each winged on one side

SEEDS: 1 per nutlet

LIGHT: Sun or partial shade

SOIL: Fertile soil with good organic content

MOISTURE: Moist

A. platanoides

pseudoplatanus

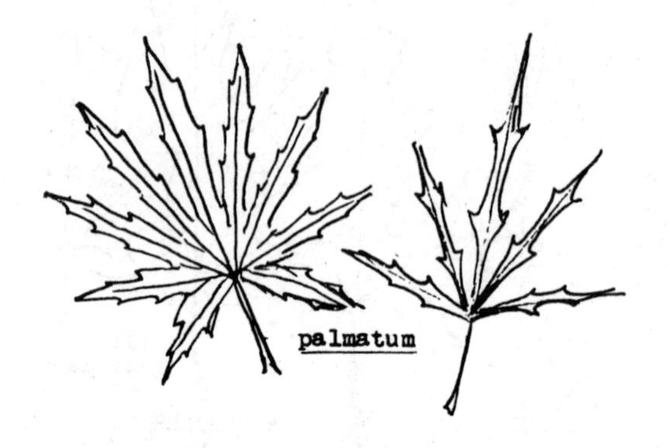

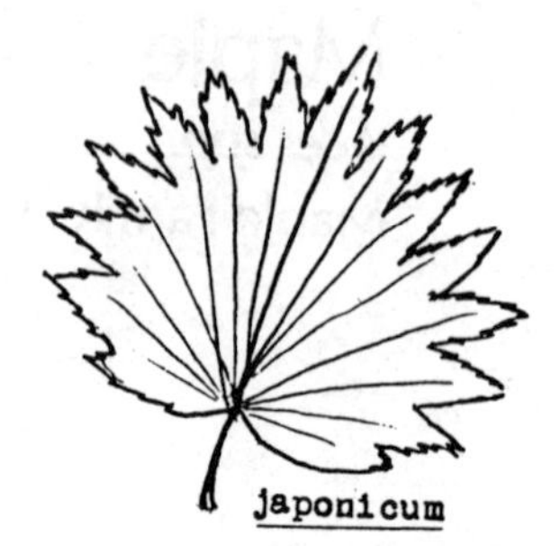

SEASONAL ASPECT: Flowers when they occur before the leaves and autumn foliage in Red and Sugar Maples

ZONE: see below

USES: Trees as background, shade, screening, framing or freestanding. Shrubs; occasional, interest, hedge.

CARE: None

PROPAGATION BY: Seedlings

NATIVE OF: e. United States, or as noted below

OTHER: Most of the species in the following list are cultivated in several to many different varieties.

Native Tree Species:

Red, *A. rubrum*; flowers before the leaves and each flower on a definite stalk.

Silver, *A. saccharinum*; flowers before the leaves and each flower sessile.

Sugar, *A. saccharum*; Flowers with the leaves.

Southern Sugar, *A. saccharum* var. *floridanum (A. barbatum)*; flowers as above; a somewhat smaller tree than above; autumn colors superb.

Introduced Tree Species:

Oregon, *A. macrophylla*; leaves lobed to middle or below; lower surface green.

Norway, *A. platanoides*; leaf lobes shallow; lower surface green.

Sycamore, *A. pseudoplatanus*; Europe; lower surface of leaves gray to whitish.

Introduced Shrub Species:

Hedge, *A. campestre*; leaf lobes shallow and obtuse.

Fullmoon, *A. japonicum*; leaf lobes shallow and acute.

Japanese, *A. palmatum*; leaves lobed to middle or below.

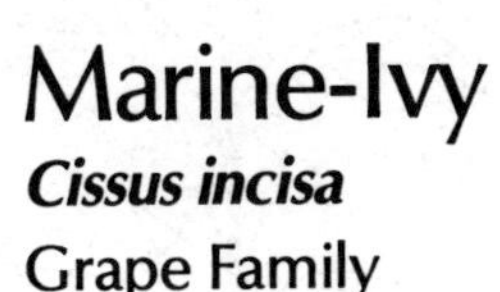

Marine-Ivy

Cissus incisa

Grape Family

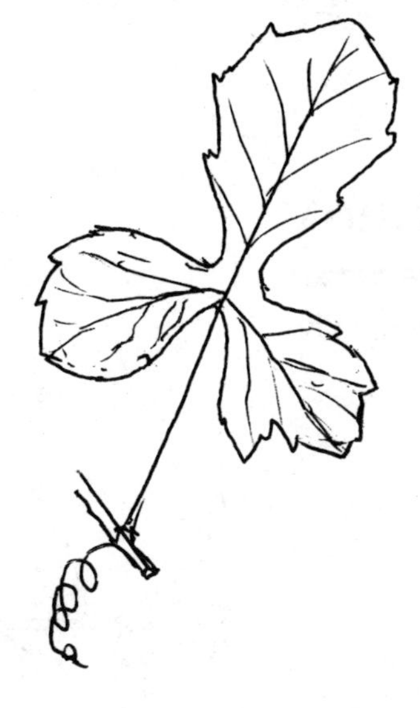

GROWTH HABIT: Deciduous, woody vine climbing by unbranched tendrils
LEAVES: Thickish, 3-parted or 3-foliate, divisions or leaflets toothed or lobed
FLOWERS: Flower cluster terminal, flowers very small
FRUITS: 1/4″ wide black berry
SEEDS: Few
LIGHT: Partial or full sun
SOIL: Tolerant
MOISTURE: Moist
SEASONAL ASPECT: None
ZONE: 1, 2
USES: Masonry, fence, trellis, arbor
CARE: None
PROPAGATION BY: Cuttings or seedlings
NATIVE OF: Florida-Missouri west and south

Marlberry

Ardisia japonica

Myrsine Family

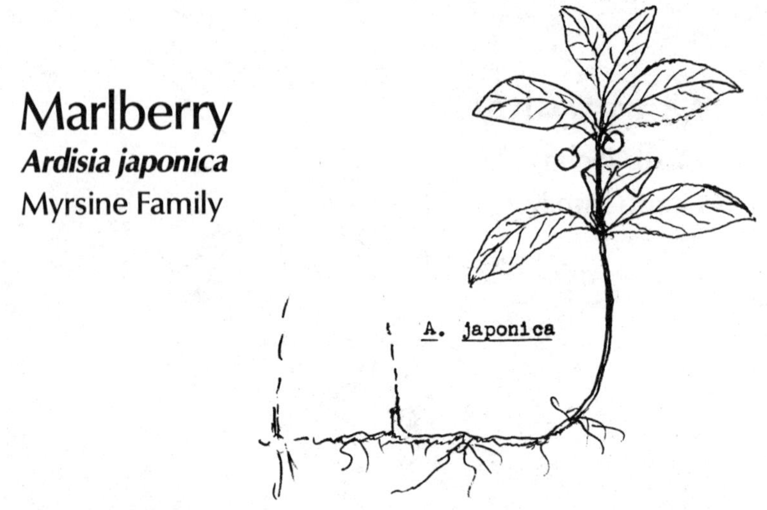

GROWTH HABIT: Low, creeping evergreen to 8″ high but not dense ground cover

LEAVES: Alternate, oblong to elliptic, margin shallowly toothed or wavey, to 2 1/2″ long

FLOWERS: White, rather small and not numerous

FRUITS: Reddish

SEEDS: 1

LIGHT: Partial shade

SOIL: Fertile

MOISTURE: Moist

SEASONAL ASPECT: Greenery for winter

ZONE: 1, 2, 3

USES: Thin ground cover, woodland planting

CARE: None

PROPAGATION BY: Colony Division

NATIVE OF: Japan

OTHER: *A. crenata*; an ornamental pot shrub in our climate with a few short branches, white flowers and red or white berry-like fruits, mistakenly referred to as Christmas Berry or Toyon (*Heteromeles*).

Marsh-Elder

Iva frutescens

Composite Family

GROWTH HABIT: Tardily decidous shrub to 6′ high
LEAVES: Opposite, elliptic, to 3″ long, toothed and hairy on both surfaces
FLOWERS: Very small, crowded into small heads surrounded by a few green bracts
FRUITS: Achene
SEEDS: 1
LIGHT: Full sun
SOIL: Coastal dunes, salt marshes and ditch banks
MOISTURE: Moist to wet
SEASONAL ASPECT: None
ZONE: 1
USES: Only for coastal landscaping
CARE: None
PROPAGATION BY: Seeds
NATIVE OF: s.e. United States
OTHER: *I. imbricata*; s.e. United States; A smaller plant with smaller and alternate leaves.

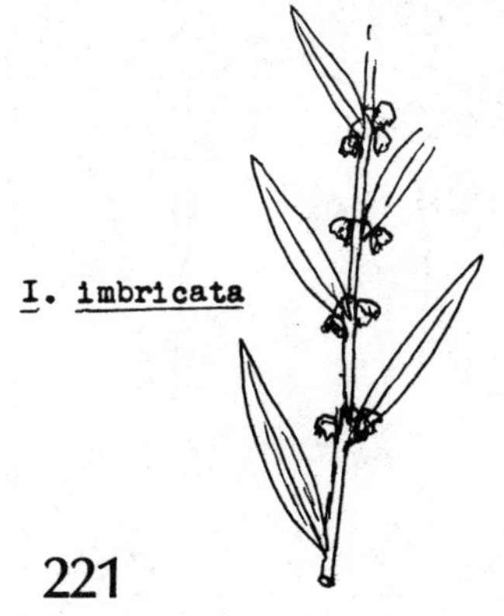

Marsh Pennywort

Hydrocotyle verticillata

Parsley Family

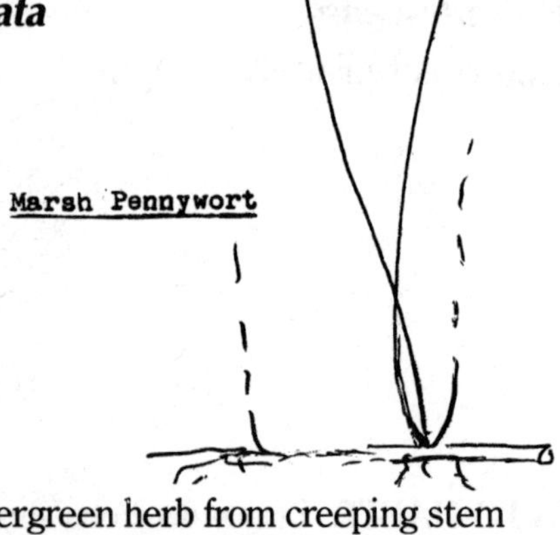

GROWTH HABIT: Low evergreen herb from creeping stem
LEAVES: Circular with stalk attached to middle
FLOWERS: Small, white and borne on separate stem
FRUITS: Paired nut-like structures
SEEDS: 2
LIGHT: Full or partial sun
SOIL: Sandy or silty
MOISTURE: Wet or moist
SEASONAL ASPECT: None
ZONE: 1, 2
USES: A ground cover for low, wettish place
CARE: None
PROPAGATION BY: Seeds and rooted stems
NATIVE OF: s.e. United States
OTHER: *Centella asiatica,* a somewhat similar plant often growing with above.

Matrimony-Vine, Box-Thorn

Lycium halimifolium

Nightshade Family

GROWTH HABIT: Tardily deciduous shrub, spreading by arching, spiny branches to 10′ long that root on contact with the ground

LEAVES: Alternate, often clustered, mostly lanceolate, to 2″ long and dull green

FLOWERS: Small, lilac-purple

FRUITS: 1/2″ long and scarlet to orange-red

SEEDS: Several

LIGHT: Partial shade

SOIL: Poor, unproductive types

MOISTURE: Dry

SEASONAL ASPECT: Autumn fruits

ZONE: 1, 2

USES: Bank, background, barrier

CARE: Some support or pruning necessary, lest it form a tangle; prone to spread into the wild

PROPAGATION BY: Seeds

NATIVE OF: s.e. Europe

OTHER: *L. chinense*; w. Asia; larger leaves, flowers and fruits; fewer thorns.

L. carolinianum; sandy shell beaches, s.e. United States, rare and in need of protection by cultivation; zone 3.

Mimosa, Silk-Tree

Albizia julibrissin

Legume Family (Pulse, Pea)

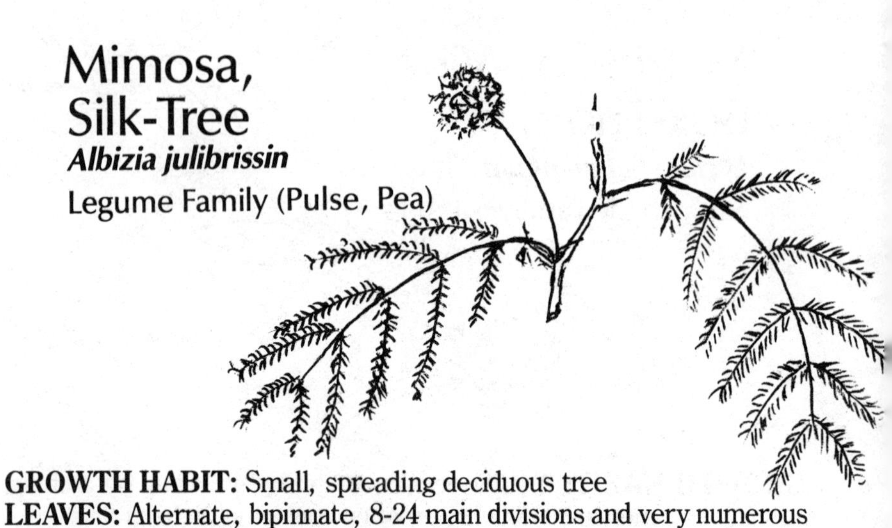

GROWTH HABIT: Small, spreading deciduous tree
LEAVES: Alternate, bipinnate, 8-24 main divisions and very numerous unequal sided 1/4″ long leaflets
FLOWERS: Pink, small and in globose heads
FRUITS: 3-5″ long papery pod
SEEDS: Several
LIGHT: Full or nearly full sun
SOIL: Fertile
MOISTURE: Moist
SEASONAL ASPECT: Flowers in early summer
ZONE: 1, 2, 3
USES: Freestanding, street, sidewalk
CARE: None
PROPAGATION BY: Seedlings
NATIVE OF: Iran to Japan

Minnie-Bush

Menziesia pilosa

Heath Family

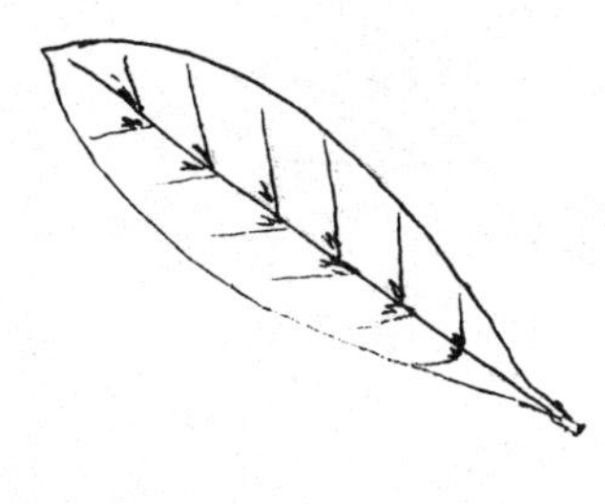

GROWTH HABIT: Deciduous shrub to 5′ high
LEAVES: Thin, elliptic, pubescent and with scale-like structures in the vein axes beneath
FLOWERS: Vase-shaped, greenish to pinkish and 1/4″ or more long, petal lobes shallow; spring
FRUITS: 4-celled capsule
SEEDS: Several
LIGHT: Partial shade
SOIL: Acid type rich in humus
MOISTURE: Moist to wet
SEASONAL ASPECT: None
ZONE: 2, 3
USES: Specimen, in a shrub border
CARE: None
PROPAGATION BY: Seedlings, transplants
NATIVE OF: Mountain province of s.e. United States

Mistletoe

Phoradendron serotinum
(P. flavescens)
Mistletoe Family

GROWTH HABIT: Evergreen shrub, partially parasitic on branches and trunks of some hardwood trees and occasionally shrubs
LEAVES: Opposite, thick, entire, about 1/2″ long and more-or-less orbicular
FLOWERS: Small, green, no petals, and in spikes; male and female on separate plants
FRUITS: White, globose
SEEDS: 2, 3
LIGHT: Partial sun, in summer
SOIL: —
MOISTURE: —
SEASONAL ASPECT: Green leaves, stems and white fruits in winter
ZONE: 1, 2, 3
USES: Christmas decoration
CARE: None
PROPAGATION BY: Seeds which have a sticky pulp and easily adhere to branches
NATIVE OF: e. United States
OTHER: Cultivated by slitting the tender bark of some hardwood (Oak, Elm, Sweet Gum, etc.) and rubbing in the seeds in late winter; fruits poisonous.

Mock-Orange

Philadelphus coronarius

Saxifrage Family

GROWTH HABIT: Deciduous, often arching shrubs to several feet high
LEAVES: Opposite, to 3″ or more long, slightly toothed and with slight pubescence beneath
FLOWERS: Usually white, 1 1/2″ across, fragrant and borne in small clusters
FRUITS: Capsule
SEEDS: Very small
LIGHT: Half shade, or more
SOIL: Fertile loam
MOISTURE: Moist
SEASONAL ASPECT: Late spring flowers
ZONE: 1, 2, 3
USES: Occasional flowering shrub, interest, undershrub in wooded areas
CARE: None
PROPAGATION BY: Cuttings and shoots
NATIVE OF: Europe; sometimes escaped here.
OTHER: *P. inodorus*; native from Pennsylvania, Georgia and Alabama confusingly similar.
P. purpurascens; w. China; to 10′ high, white, single, 1″ wide and very fragrant; flowers in purple calyx.
P. X splendens (*P. grandiflorus X P. gordonianus*); flowers white, single 1 1/2″ wide; branches arching; a good specimen plant.

Mock Orange

Poncirus trifoliata

Rue Family

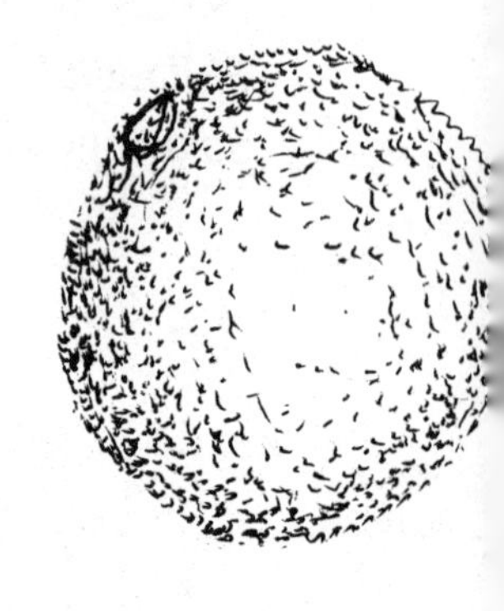

GROWTH HABIT: Deciduous shrub with stout, green thorns, twigs and branches
LEAVES: Trifoliate with smooth, entire leaflets and winged petioles
FLOWERS: Axillary, white and fragrant
FRUITS: Orange-like but about golf-ball size and very fragrant
SEEDS: Many
LIGHT: Full or three-fourth sun
SOIL: Porous, moderately fertile
MOISTURE: Moist
SEASONAL ASPECT: Flowering time, fruiting time and winter when green twigs are conspicuous
ZONE: 1, 2, 3
USES: Interest shrub or in barrier plantings
CARE: Some pruning may be desirable depending on use
PROPAGATION BY: Seeds and seedlings
NATIVE OF:
OTHER: Grated outer peel has been recommended for use in making "lemon custards".

Moltkia

Moltkia petraea

Borage Family

GROWTH HABIT: Semievergreen grayish shrub to 1 1/2′ high
LEAVES: Gray-green, and 1″ long
FLOWERS: Pinkish-purple turning bluish, 1/2″ long, in clusters
FRUITS: Nutlets
SEEDS: 1-4
LIGHT: Full sun
SOIL: Thin
MOISTURE: Dry
SEASONAL ASPECT: Summer flowers
ZONE: 1, 2
USES: Rock garden
CARE: None
PROPAGATION BY: Seeds
NATIVE OF: s.e. Europe
OTHER: Has general appearance of Lavender.

Mondo-Grass

Ophiopogon jaburan

Lily Family

GROWTH HABIT: Grass-leaved evergreen, 10-18″ high and clump-forming from short heavy rhizomes
LEAVES: 1/2 to 3/4″ wide arising from ground level
FLOWERS: White or bluish and drooping on pedicels longer than flowers
FRUITS: Dark blue
SEEDS: 1 to few
LIGHT: Tolerant of shade
SOIL: Fertile loam
MOISTURE: Moist
SEASONAL ASPECT: None
ZONE: 1, 2, 3
USES: Interest clump, border, edging, sidewalk planting
CARE: None
PROPAGATION BY: Clump division
NATIVE OF: Japan
OTHER: A striped-leaf form is known.

Monkey-Puzzle

Araucaria araucana

Araucaria Family

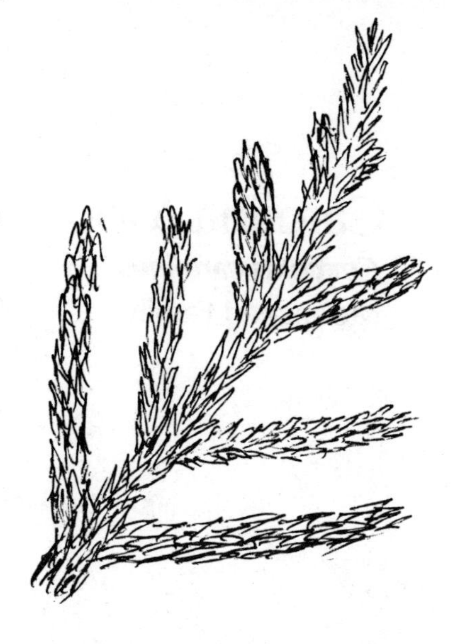

GROWTH HABIT: Tall conical evergreen tree with upward curving branches

LEAVES: 1 1/2″ long, lanceolate, sharp-pointed, stiff, closely imbricated on twigs and branches and persisting for several years

FLOWERS: Male as 3-4″ long catkins; female head-like and becoming woody cones

FRUITS: 6″ long globose cones

SEEDS: 1 per scale

LIGHT: Full sun

SOIL: Fertile loam

MOISTURE: Moist

SEASONAL ASPECT: None

ZONE: 1, 2

USES: Interest or accent

CARE: None

PROPAGATION BY: Seedlings

NATIVE OF: Chile

OTHER: *A. excelsa,* Norfolk Island Pine, S. Pacific, grown as pot plant indoors.

Carolina Moonseed

Cocculus carolinus

Moonseed Family

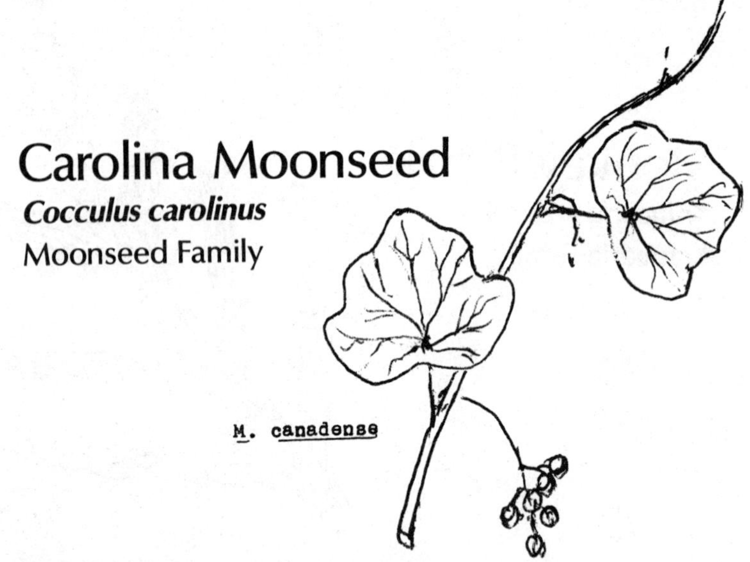

GROWTH HABIT: Deciduous woody twining vine
LEAVES: Alternate, broadly ovate, entire or shallowly lobed
FLOWERS: Male and female on separate plants, very small, greenish
FRUITS: Bright red berries about 1/4″ wide
SEEDS: 1
LIGHT: Full sun for lots of berries
SOIL: Almost any type
MOISTURE: Dry
SEASONAL ASPECT: Fall fruits
ZONE: 1, 2
USES: Fence, arbor, trellis
CARE: None
PROPAGATION BY: Seeds
NATIVE OF: s.e. United States
OTHER: Moonseed, *Menispermum canadense*; e. United States; deciduous twining vine with palmately lobed leaves and black fruits.

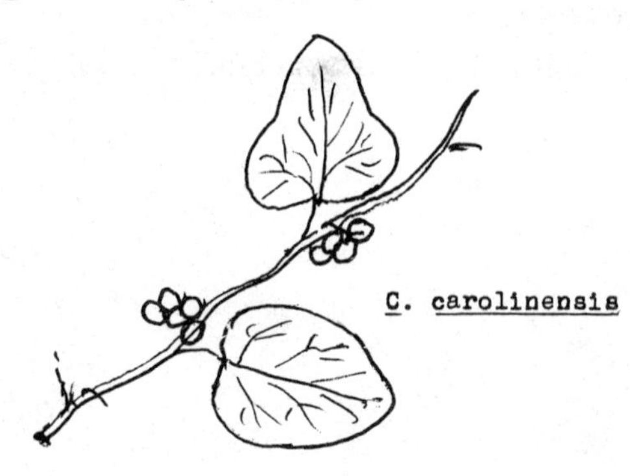

Mountain Ash

Sorbus spp.

Rose Family

GROWTH HABIT: Small to medium deciduous trees, or large shrub
LEAVES: Alternate, pinnate with from 9-17 leaflets that are lanceolate, toothed and to 3″ long
FLOWERS: White, about 1/4″ wide and in dense flat-topped clusters to 6″ wide
FRUITS: Bright red and shining, 1/4″ wide, long lasting and ornamental
SEEDS: 1-2
LIGHT: About 3/4 sun
SOIL: Fertile
MOISTURE: Moist
SEASONAL ASPECT: Bright red fruits in fall and winter
ZONE: 2, 3
USES: Freestanding, interest, background
CARE: None
PROPAGATION BY: Seedlings
NATIVE OF: See below
OTHER: American Mountain-Ash, S. *americana*; shrub or small tree.
European Mountain-Ash, S. *aucuparia*; somewhat larger tree; leaflets pubescent beneath.

Mountain Laurel, Ivy

Kalmia latifolia

Heath Family

GROWTH HABIT: Large, much-branched, evergreen shrub
LEAVES: Alternate, thick, elliptic, entire and to 4″ long
FLOWERS: 1/2″ or more wide, pink, purplish or white, saucer shaped but with 5 shallow lobes and 10 anther pockets
FRUITS: 1/4″ long capsules
SEEDS: Very small
LIGHT: Partial shade
SOIL: Most any well drained acid type
MOISTURE: Moist but not wet
SEASONAL ASPECT: Spring flowers
ZONE: 1, 2, 3
USES: Occasional flowering shrub, hedge, specimen
CARE: Some pruning
PROPAGATION BY: Seedlings
NATIVE OF: e. United States
OTHER: The most successful transplanting involves removal of entire top about 6″ above ground level. ″Breaking Ivy“ is still practiced by mountain folk and is the breaking off of the leafy twigs to sell by the pound to big city florists.

Lambkill, *K. angustifolia*; smaller leaves in 3′s; dependent on wet acid soil.

Mulberry

Morus spp.

Mulberry Family

GROWTH HABIT: Spreading deciduous trees

LEAVES: Alternate, ovate, toothed, some variously lobed, mostly rough above

FLOWERS: Small and on short pendant spikes; male and female on separate trees

FRUITS: Many very small juicy drupelets on a central axis; edible

SEEDS: 1 per drupelet

LIGHT: Full or partial sun

SOIL: Tolerant

MOISTURE: Moist

SEASONAL ASPECT: None

ZONE: 1, 2, 3

USES: Freestanding, specimen tree, street planting

CARE: None

PROPAGATION BY: Seedlings and cuttngs

NATIVE OF: See below

OTHER: Red Mulberry, *M. rubra,* e. United States; medium to large tree with broad crown and leaves that are rough above, pubescent beneath and barely heart-shaped at base.

Black Mulberry, *M. nigra*; Asia; small tree with leaves that are pubescent beneath and deeply heart-shaped at base.

White Mulberry, *M. alba,* China; small tree with smooth glossy leaves; one variety has drooping branches.

Paper Mulberry

Broussonetia papyrifera

Mulberry Family

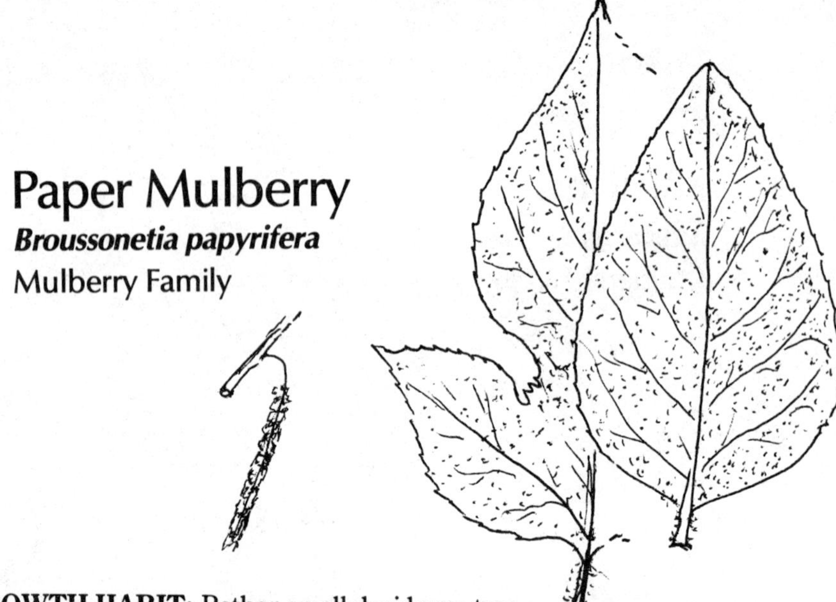

GROWTH HABIT: Rather small deciduous tree
LEAVES: Opposite and alternate, ovate, toothed and densely grayish-pubescent, often variously lobed
FLOWERS: Male and female on different trees, very small and in drooping spikes; female trees rare
FRUITS: Rarely developed
SEEDS: —
LIGHT: Sun
SOIL: Most any
MOISTURE: Dry
SEASONAL ASPECT: None
ZONE: 1, 2, and 3
USES: As a specimen tree or wherever other trees are unlikely to prosper
CARE: None
PROPAGATION BY: Sprouts and cuttings
NATIVE OF: China and Japan
OTHER: An extremely hardy plant, often seen in waste areas, railroad rights-of-way, etc., where it spreads by root sprouts.

Muscadine, Bullace, Scuppernong

Vitis rotundifolia

Grape Family

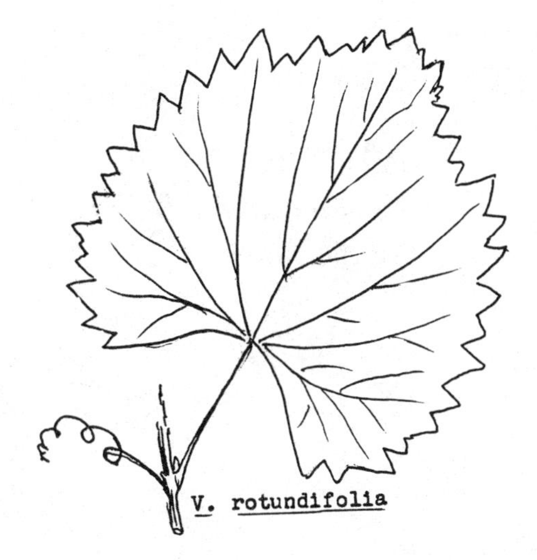

GROWTH HABIT: Deciduous woody tendril-bearing vine with shreading bark
LEAVES: 2-4″ wide, cordate, coarsely toothed and glabrous
FLOWERS: Small, greenish, unisexual and 5-parted, the 5 petals falling together
FRUITS: Black or amber berries to 1″ wide
SEEDS: 1-3 (5)
LIGHT: Full sun
SOIL: Tolerant
MOISTURE: Moist
SEASONAL ASPECT: Fruits in late summer or early fall
ZONE: 1, 2, 3
USES: Arbor, trellis, fence, fruit
CARE: Pruning may be desirable
PROPAGATION BY: Cuttings
NATIVE OF: e. United States
OTHER: Other native and introduced species are cultivated for their fruits but require pruning, spraying, fertilizing, and cultivation.

V. rotundifolia is so hardy as to reward its keeper with screening, shade and fruit for only the space in which to grow.

Scarlet-leaved Vine, *V. amurensis*; an introduction from Manchuria; the large 3- to 5-lobed leaves are prettily colored rose to crimson; autumn colors vivid.

Japanese Crimson Glory Vine, *V. coignetiae*; a robust climber with large green leaves; rusty beneath that change to amber, pink, rose and finally crimson in autumn; use as cover-up for unsightly walls, buildings or other structures.

Myrtle

Myrtus communis

Myrtle Family

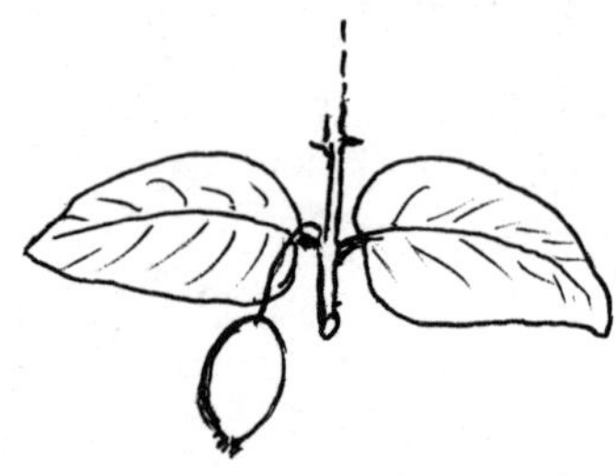

GROWTH HABIT: Compact evergreen shrub with pleasant fragrance
LEAVES: Opposite, entire, ovate, smooth and shiny and to 1 1/2″ long
FLOWERS: White or tinted, 1/2″ wide and on stalks
FRUITS: Dark blue (or white) berries
SEEDS: Several
LIGHT: Full sun
SOIL: Fertile loam
MOISTURE: Moist
SEASONAL ASPECT: None
ZONE: 1, 2
USES: Interest, specimen
CARE: Cold sensitive but usually sprouts back
PROPAGATION BY: Cuttings and seedlings
NATIVE OF: w. Asia and the Mediterranean region
OTHER: This is the classic Myrtle and is cultivated in many varieties.

Nandina

Nandina domestica

Barberry Family

GROWTH HABIT: Evergreen shrub with little or unbranched stems
LEAVES: Alternate, clusters near stem tips, 2 or 3 X compound with many leaflets 1/2-1 1/2″ long
FLOWERS: Small, white and in large terminal panicles
FRUITS: Bright red berries about 1/4″ wide
SEEDS: Few
LIGHT: Sun or shade
SOIL: Fertile loam
MOISTURE: Moist
SEASONAL ASPECT: Red berries in fall
ZONE: 1, 2, 3
USES: Singly or in groups of 3 or as facing for larger shrubs or wall
CARE: Remove older canes; cutting them off at different heights
PROPAGATION BY: Seedlings and division of clumps
NATIVE OF: Japan
OTHER: A dwarf variety is cultivated, also one with white berries.

Nestronia

Nestronia umbellata

Sandalwood Family

GROWTH HABIT: Deciduous shrub to 2′ high
LEAVES: Alternate, broadly elliptic, entire and to 2 1/2″ long
FLOWERS: Male and female usually on different plants; small and greenish
FRUITS: Yellow-green and 1/2″ long
SEEDS: Few
LIGHT: Half shade
SOIL: Tolerant to most types
MOISTURE: Moist
SEASONAL ASPECT: None
ZONE: 2
USES: Interest
CARE: Keep among hardwood species and if it can be grown adds interest to the natural effect
PROPAGATION BY: Seeds
NATIVE OF: s.e. United States
OTHER: Probably parasitic on the roots of deciduous trees and shrubs, doing no apparent harm.

New Jersey Tea

Ceanothus americanus

Buckthorn Family

GROWTH HABIT: Tardily deciduous shrub to 3′ high
LEAVES: Alternate, dull green, finely toothed, 3-veined and 1/2 - 3″ long
FLOWERS: Small, white and in 1-2″ long, elevated clusters
FRUITS: 3-lobed capsule-like structure
SEEDS: 3
LIGHT: Partial shade
SOIL: Rather sterile, porous type
MOISTURE: Dry to moist
SEASONAL ASPECT: Early summer flowers
ZONE: 1, 2, 3
USES: Occasional shrub, interest
CARE: None
PROPOGATION BY: Seedlings
NATIVE OF: e. United States
OTHER: During the Revolution this was thought to be a possible source of tea. Other species are cultivated, mostly in the west.

Ninebark

Physocarpus opulifolius

Rose Family

GROWTH HABIT: Strong, arching, deciduous shrub to 10′ high
LEAVES: Ovate in outline; mostly 3-lobed, toothed along the margin and 1-3″ long
FLOWERS: White to pinkish or yellowish in 2″ wide dense clusters and on leafy shoots of current season
FRUITS: 3-5 small, dry pods (follicles) per flower
SEEDS: Few per pod
LIGHT: Full sun for best blossoms
SOIL: Fertile silty or sandy loam
MOISTURE: Moist
SEASONAL ASPECT: Summer flowers
ZONE: 1, 2, 3
USES: Occasional flowering shrub
CARE: Pruning for desired size and shape
PROPAGATION BY: Cuttings
NATIVE OF: E. North America

Nolina

Nolina georgiana

Lily Family

GROWTH HABIT: Evergreen rosette former, from bulbous base

LEAVES: Narrowly linear, to 2′ long and forming rather dense tuft or basal rosette

FLOWERS: White, less than 1/4″ long but numerous on large panicle to 4′ high

FRUITS: Papery 3-celled capsule

SEEDS: Brownish

LIGHT: Full sun

SOIL: Sandy

MOISTURE: Dry

SEASONAL ASPECT: Late spring flowers

ZONE: 1, 2

USES: Interest, specimen

CARE: None

PROPAGATION BY: Seeds or plants

NATIVE OF: s.e. United States

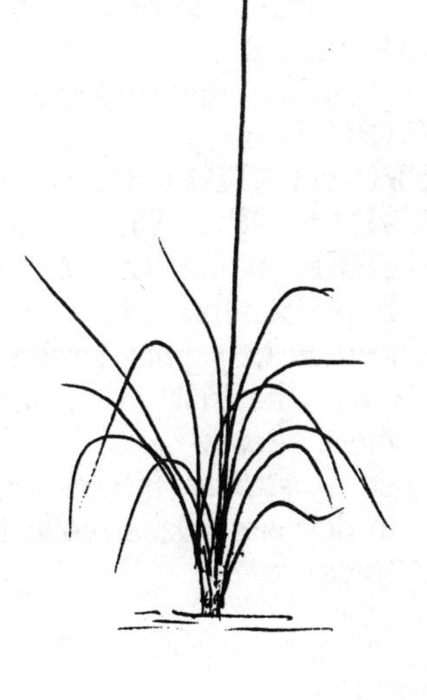

Oak, White, Chestnut, Post

Quercus spp.

Beech Family

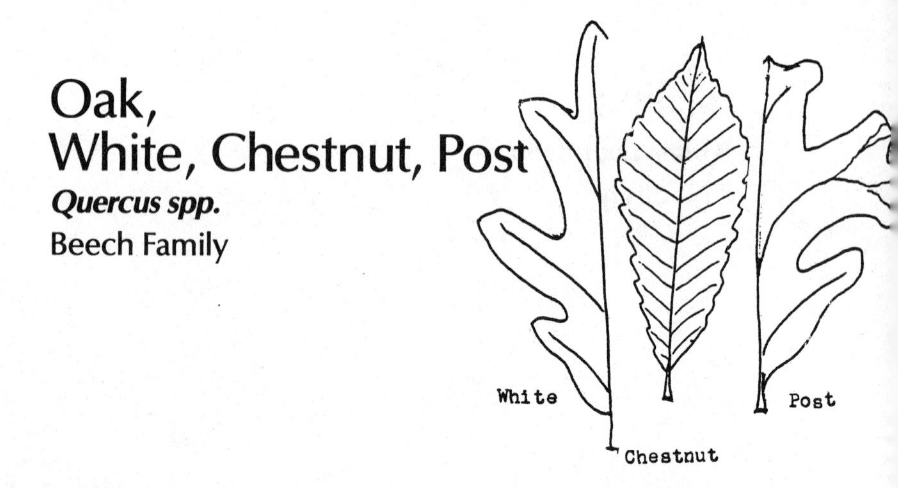

GROWTH HABIT: Large deciduous trees with spreading crowns and acorns that are cream colored inside and germinate in the fall
LEAVES: Alternate 2 1/2 - 7″ long with rounded lobe tips and sinuses that are rounded in the bottom; variable within the species
FLOWERS: Male in thin drooping catkins; female small, brownish and inconspicuous
FRUITS: Acorns in cups; 3/4 - 2″ long
SEEDS: 1 per acorn
LIGHT: At least partial sun
SOIL: Various loams
MOISTURE: Moist to dry
SEASONAL ASPECT: White Oak foliage is brightly colored in fall
ZONE: 1, 2, 3
USES: Framing, background, shade, street and drive
CARE: None
PROPAGATION BY: Seedlings
NATIVE OF: e. United States
OTHER: White, *Q. alba*; a long time favorite hardy ornamental with bright autumn colors.
Chestnut, *Q. prinus*; large fruited Oak developing some autumn colors.
Swamp Chestnut, *Q. michauxii*; similar to above but requires a more moist site.
Post, *Q. stellata*; interesting leaf-shade and yellow autumn color. Plant in dry site. Oaks, collectively, are our most important ornamental trees.

Oak,
Scarlet, Pin, Red, Black, Spanish, Cherrybark, Turkey

Quercus spp.

Beech Family

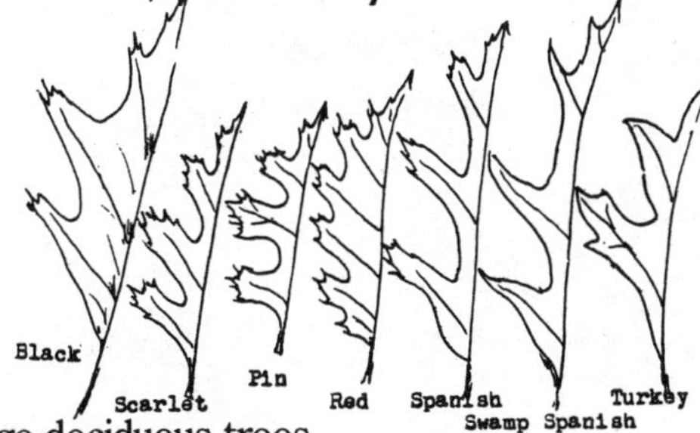

GROWTH HABIT: Medium to large deciduous trees

LEAVES: Alternate, 3-10″ long, with bristle-tipped lobes and sinuses of varying depths, variable within the species

FLOWERS: Male in thin drooping catkins before the leaves; female small, brownish and inconspicuous

FRUITS: Acorns in cups

SEEDS: Orange inside and germinating in spring

LIGHT: Half sun or more

SOIL: Fertile loam

MOISTURE: Moist

SEASONAL ASPECT: See Scarlet, Red and Turkey Oaks below

ZONE: See below

USES: Wherever deciduous trees are desired

CARE: None

PROPAGATION BY: Seedlings

NATIVE OF: e. United States

OTHER: Scarlet, *Q. coccinea*; zones 1, 2, 3; acorn with 2 or 3 faint grooves around point; leaves deeply cut, smooth beneath and colorful in autumn.

Pin, *Q. palustris*; zone 2; somewhat smaller leaves than above; acorn cup shallow; dead leaves in fall long persisting.

Red, *Q. rubra*; zones 2, 3; leaves cut only about half way to mid vein; good autumn color.

Black, *Q. velutina*; zones 1, 2, 3; leaves with tufts of hairs in the vein axes beneath.

Spanish or Southern Red, *Q. falcata*; zones 1, 2, 3; leaves densely yellowish-brown pubescent beneath and mostly with bell-shaped bases.

Swamp Spanish, Cherrybark; *Q. pagodaefolia*; zones 1, 2; leaves not bell-shaped at base, grayish pubescent beneath and deeply cut.

Turkey, *Q. laevis*; zones 1, 2; small tree and dry sandy sites with nice autumn colors. Other natives Oaks are sometimes used.

Oak,
Water, Willow, Laurel

Quercus spp.

Beech Family

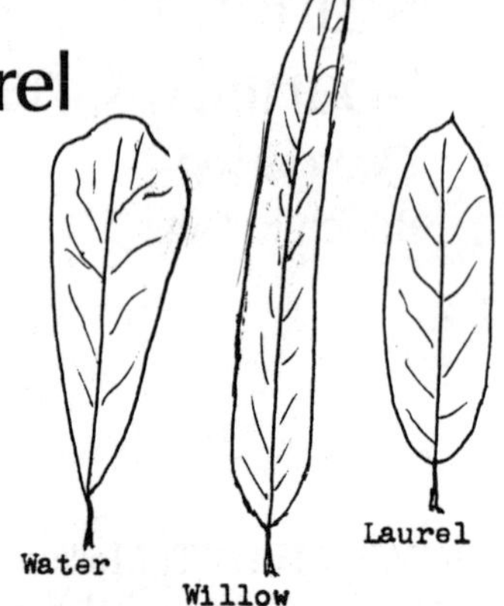

GROWTH HABIT: Widely spreading deciduous trees
LEAVES: Alternate, small, narrow, mostly unlobed or shallowly so
FLOWERS: Male as thin drooping catkins; female small, brownish and inconspicuous
FRUITS: Acorns small, orange within and germinate in spring
SEEDS: 1 per fruit
LIGHT: Full or lots of sun
SOIL: Fertile silt or sandy loam
MOISTURE: Moist
SEASONAL ASPECT: None
ZONE: 1, 2, 3
USES: Freestanding, framing, background, shade, street
CARE: None
PROPAGATION BY: Seedlings
NATIVE OF: e. United States
OTHER: Water Oak, *Q. nigra*; leaves broadest toward tips and sometimes 3-lobed, variable in seedlings.

Willow Oak, *Q. phellos*; leaves long lanceolate thin and generally drooping.

Laurel or Darlington Oak, *Q. laurifolia*; leaves oblong, thickish and tardily deciduous.

Live Oak

Quercus virginiana

Beech Family

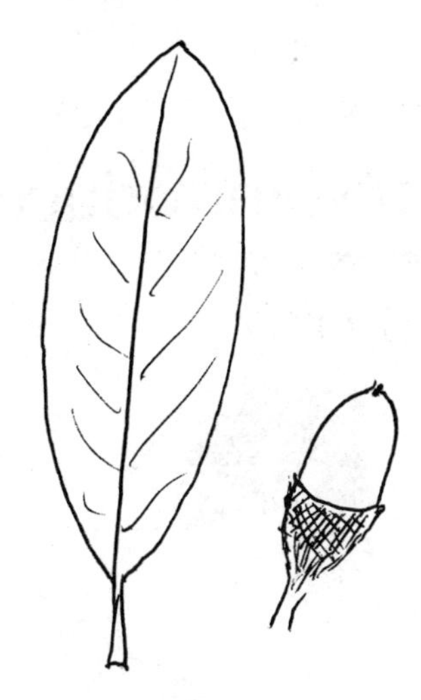

GROWTH HABIT: Large, very widely spreading evergreen tree with very short trunk
LEAVES: Thick, 1 1/2 - 4″ long, widely elliptic, with rolled-under margins, dark green above and usually grayish beneath
FLOWERS: Male in thin pendulous catkins; female small, brownish and inconspicuous
FRUITS: Acorns 1″ long
SEEDS: Orange colored inside, germinate in spring
LIGHT: Full or lots of sun
SOIL: Sandy loam
MOISTURE: Moist
SEASONAL ASPECT: None
ZONE: 1, 2
USES: This tree will eventually claim lots of space and exclude light so that little grows beneath it.
CARE: None
PROPAGATION BY: Seedlings
NATIVE OF: s.e. United States

Oaks, Introduced

Quercus spp.

Beech Family

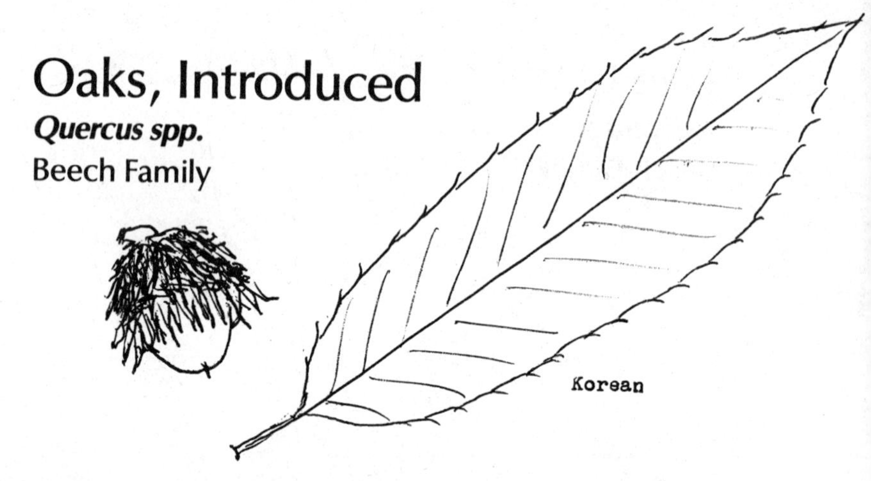

Japanese Red, *Q. acuta*; small evergreen tree; leaves obovate, toothed and to 5″ long; acorn less than 1/2″ long, in shallow cup; cup made up of 3-5 thin concentric or nested scales.

Korean or Sawtooth Oak, *Q. acutissima*; medium-sized deciduous tree; leaves to 6″, oblong and with many shallow bristle-tipped teeth; acorn and cup 1″ long; cup covering half or more of acorn; cup scales long attenuated and bristle-like, making cup somewhat bur like.

Cork Oak, *Q. suber*; s. Europe, n. Africa; small evergreen tree with deeply furrowed springy bark and rather narrow toothed leaves.

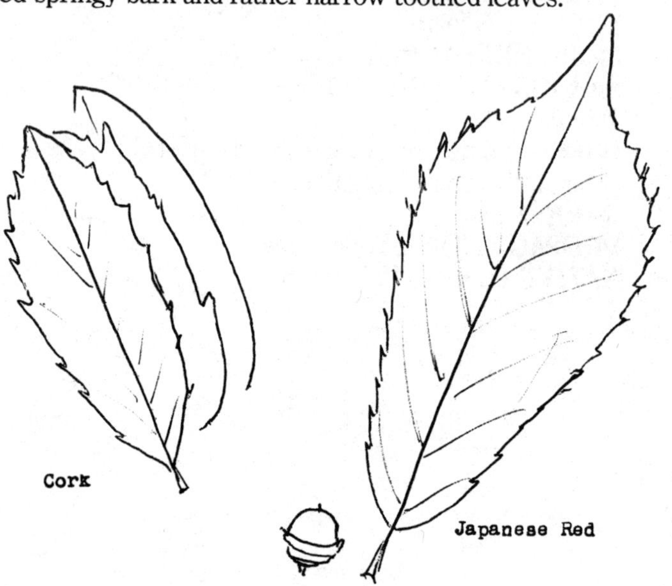

Oconee-Bells, Jackscrew-Root

Shortia galacifolia

Diapensia Family

GROWTH HABIT: Low, creeping evergreen
LEAVES: Orbicular with rounded or heart-shaped bases
FLOWERS: White, petals 5 and fringed, borne one per stalk
FRUITS: Capsule
SEEDS: Several and small
LIGHT: Shade
SOIL: Acid loam
MOISTURE: Moist
SEASONAL ASPECT: Spring flowers
ZONE: 2, 3
USES: Interest
CARE: None
PROPAGATION BY: Colony division
NATIVE OF: w. Carolinas

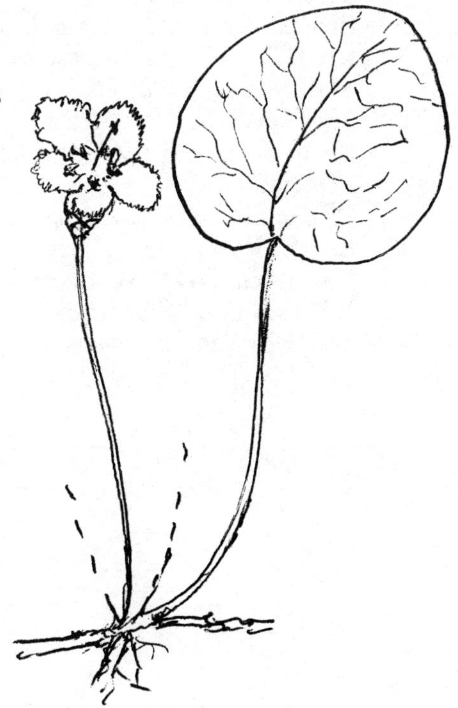

Oleander

Nerium oleander

Dogbane Family

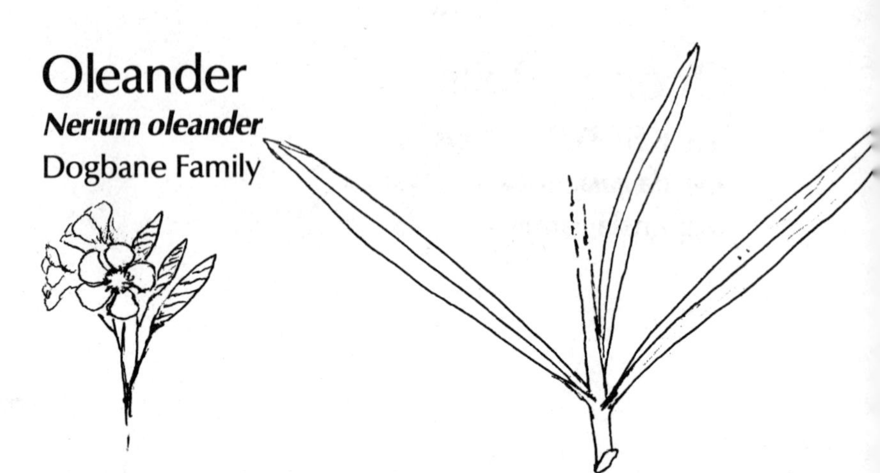

GROWTH HABIT: Rather large evergreen shrub to 15′
LEAVES: Mostly in whorls of 3, narrowly lanceolate, smooth and leathery
FLOWERS: In showy terminal clusters, white, yellow, red or purple, each flower from 1 1/2 - 3″ wide, doubled in some varieties
FRUITS: Pencil-size pods to 6″ long
SEEDS: With wisp of silky hairs at tip
LIGHT: Full sun for blossoms
SOIL: Fertile sandy loam
MOISTURE: Moist to dry
SEASONAL ASPECT: Flowers in spring, summer and early fall
ZONE: 1, 2
USES: Streets, unit arrangements, occasional shrub, screen
CARE: Root prune to keep to size
PROPAGATION BY: Seedlings and cuttings
NATIVE OF: Mediterranean regions
OTHER: Very poisonous

Opopanax

Acacia farnesiana

Legume Family

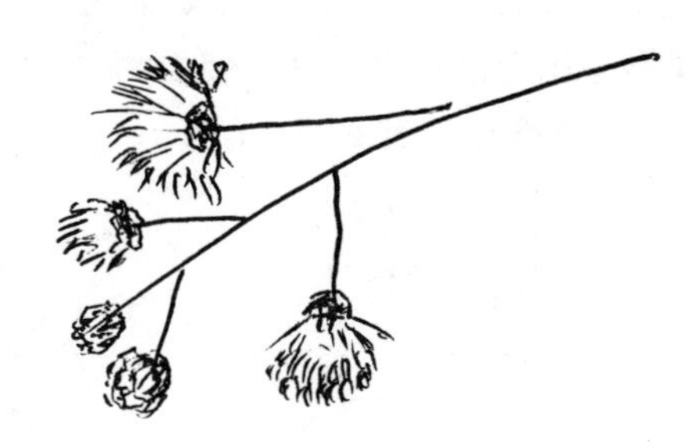

GROWTH HABIT: Widely branched, thorny deciduous shrub to 6′ high
LEAVES: Alternate, bipinnate, leaflets small, many, paired
FLOWERS: Deep yellow, small but in dense heads 1/2″ wide; fragrant
FRUITS: Pod about 2″ long
SEEDS: Several
LIGHT: Full sun
SOIL: Most any type
MOISTURE: Dry
SEASONAL ASPECT: Blooms in very early spring
ZONE: 1
USES: Occasional shrub, barrier
CARE: Remove crowded branches
PROPAGATION BY: Seeds
NATIVE OF: s.w. United States and Mexico

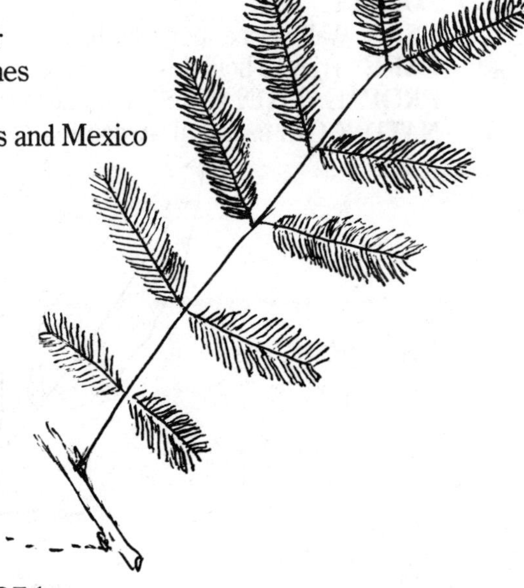

Orchid Trumpet Vine

Clytostoma callistegioides

Trumpet Creeper Family

GROWTH HABIT: Rather vigorous evergreen vine climbing by tendrils
LEAVES: Of 2 elliptic leaflets and a terminal single tendril
FLOWERS: 3″ long and pink to lilac
FRUITS: Capsule 3-5″ long and prickly
SEEDS: Many, flat and prickly
LIGHT: Full sun or partial shade
SOIL: Sandy
MOISTURE: Moist
SEASONAL ASPECT: Spring flowers
ZONE: 1
USES: Wall, fence, pergola, bank
CARE: Trim to hold desired shape
PROPAGATION BY: Seeds, cuttings
NATIVE OF: s. Brazil, Argentina

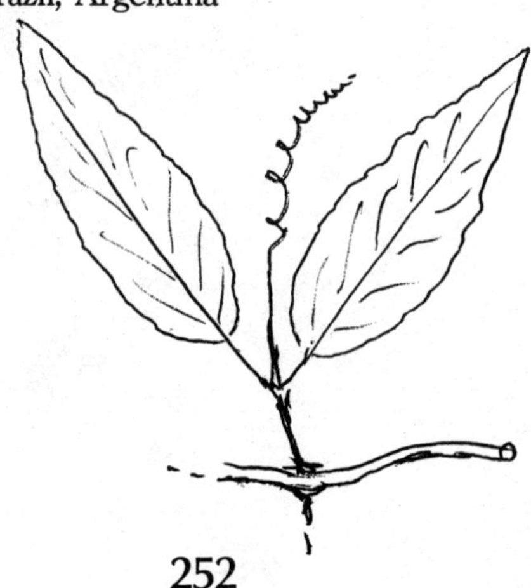

Oregon Grape-Holly

Mahonia spp.

Barberry Family

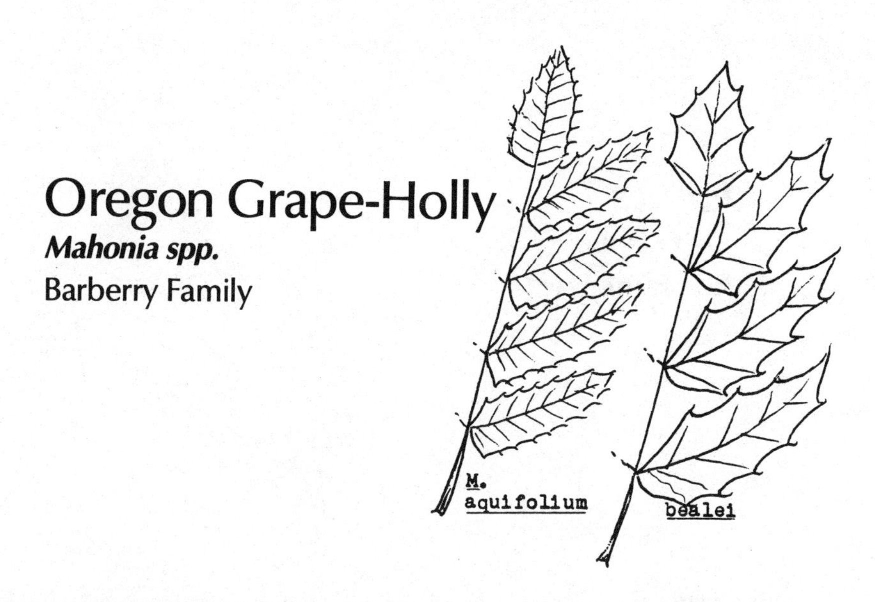

GROWTH HABIT: Little-branched evergreen shrubs to 3′
LEAVES: Alternate and clustered near stem tips, pinnate, leaflets 5-15 with spine-tipped shallow lobes or teeth
FLOWERS: Small, yellow and slightly fragrant
FRUITS: Bluish berries
SEEDS: Small
LIGHT: Partial shade
SOIL: Fertile
MOISTURE: Moist but well drained
SEASONAL ASPECT: Spring when leaves are new
ZONE: 1, 2
USES: Foundation, interest, blending
CARE: Little
PROPAGATION BY: Seeds, seedlings, clump divisions and cuttings
NATIVE OF: See below
OTHER: Oregon-Grape, *M. aquifolium*; w. United States; to 3′ high; leaflets 3-9.
Holly Mahonia, *M. bealei*; China; to 10′ high; leaflets 9-15.

Orixa

Orixa japonica

Rue Family

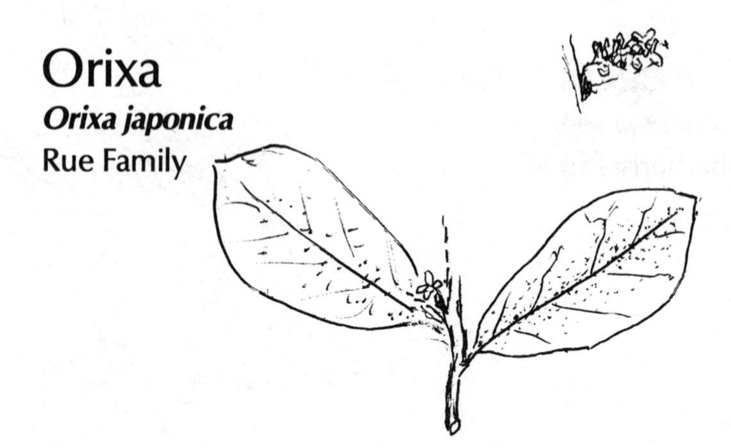

GROWTH HABIT: Deciduous shrub to 10′ high
LEAVES: Alternate, obovate, entire or nearly so, glandular dotted beneath and to 5″ long
FLOWERS: Male and female on separate plants; male in racemes, female solitary; both small, greenish and 4-parted
FRUITS: Four 2-valved sections
SEEDS: Black
LIGHT: Partial shade
SOIL: Fertile loam
MOISTURE: Moist to dry
SEASONAL ASPECT: None
ZONE: 1, 2, 3
USES: Occasional or specimen shrub
CARE: Remove crowded branches
PROPAGATION BY: Cuttings, seeds
NATIVE OF: Japan

Osage Orange

Maclura pomifera

Mulberry Family

GROWTH HABIT: Rather small deciduous tree mostly with spiny branches
LEAVES: Alternate, entire, broadly lanceolate, long pointed and with milky juice
FLOWERS: Inconspicuous, male and female on different trees
FRUITS: Orange size, green
SEEDS: Many
LIGHT: Full or partial sun
SOIL: Tolerant to most
MOISTURE: Dry or moist, even almost wet
SEASONAL ASPECT: None, or when female tree is heavily fruited
ZONE: 1, 2
USES: Interest, freestanding
CARE: None
PROPAGATION BY: Seeds or seedlings
NATIVE OF: s.e. United States
OTHER: Wood very hard, rot resistant and yellowish. Sometimes known as Bois-d'Arc Tree.

Japanese Pachysandra

Pachysandra terminalis

Box Family

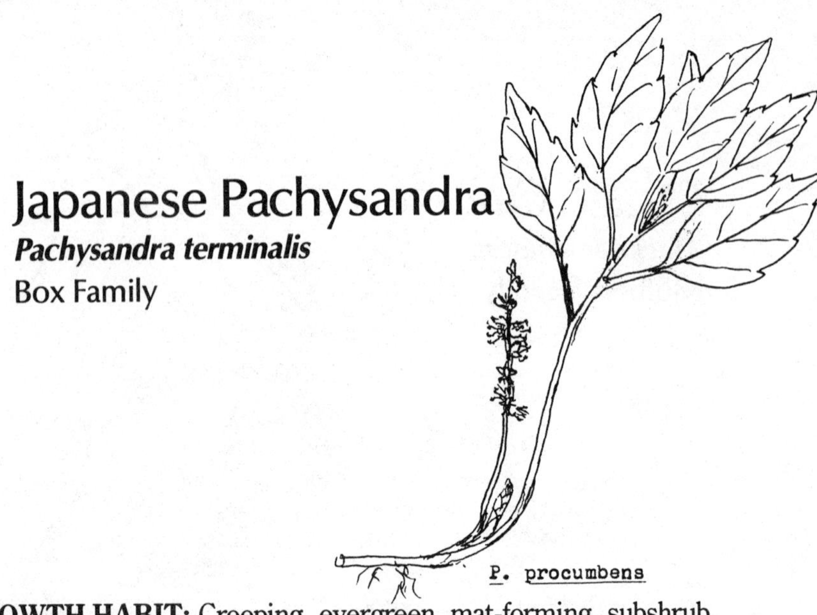

GROWTH HABIT: Creeping, evergreen, mat-forming, subshrub
LEAVES: Alternate, obovate, to 3″ long, coarsely toothed above the middle
FLOWERS: Small, white, in 1-2″ long spikes, no petals; 4 sepals; male above, female below
FRUITS: When produced a 3-horned capsule
SEEDS: 2 per cell
LIGHT: Mostly or all shade
SOIL: Good loam with humus
MOISTURE: Moist
SEASONAL ASPECT: None
ZONE: 1, 2
USES: Ground cover
CARE: None
PROPAGATION BY: Colony division, rooted stems
NATIVE OF: Japan
OTHER: *P. procumbens*; rare native species; similar but deciduous or semievergreen.

Palm

Butia capitata

Palm Family

GROWTH HABIT: Stout, usually short, semievergreen
LEAVES: Long pinnate with 25-50 pairs of leaflets with a somewhat bluish tint; petiole with marginal spines
FLOWERS: Inflorescence large and borne among the leaves; male flowers terminal
FRUITS: Yellow drupe about 1″ long
SEEDS: 1 per fruit
LIGHT: Full sun
SOIL: Fertile loam
MOISTURE: Moist
SEASONAL ASPECT: None
ZONE: 1, 2
USES: Interest, street planting, freestanding
CARE: Removal of lower leaves as they die
PROPAGATION BY: Seedlings
NATIVE OF: s. Brazil
OTHER: Six or more varieties are cultivated.

Windmill Palm

Trachycarpus fortunei

Palm Family

GROWTH HABIT: Slender evergreen shrub or small tree with coarse black or brown fibers covering trunk
LEAVES: Fan-shaped with long, rough-edged petioles
FLOWERS: Small, yellowish and in large clusters among the leaves
FRUITS: Three-angled and about 1/2″ long
SEEDS: 3
LIGHT: Full sun or partial shade
SOIL: Sandy loam
MOISTURE: Moist
SEASONAL ASPECT: None
ZONE: 1, 2
USES: Freestanding, interest or tropical effect
CARE: Little or none
PROPAGATION BY: Seedlings and stem offsets
NATIVE OF: e. Asia
OTHER: A very hardy palm

Palmetto

Sabal palmetto

Palm Family

GROWTH HABIT: Familiar unbranched evergreen tree
LEAVES: More-or-less fan-shaped but curving downward along the midrib
FLOWERS: Small, greenish and in large, long clusters among the leaves
FRUITS: Black, drupe-like, thin pulp, large seed; 1/3″ thick and globose
SEEDS: 1 per fruit
LIGHT: Full or partial sun
SOIL: Silt or silt loam
MOISTURE: Moist to wet
SEASONAL ASPECT: None
ZONE: 1, and 2 by transplant and protection
USES: Interest, street planting
CARE: Remove lower leaves as they die
PROPAGATION BY: Seedlings
NATIVE OF: Coastal North Carolina and south
OTHER: The State Tree of South Carolina

Blue Palmetto

Raphidophyllum histryx
Palm Family

GROWTH HABIT: Evergreen shrub with erect or stooping trunk to 5′
LEAVES: Fan-shaped, no fibers on the ultimate divisions, leaf stalk sharp edged, sheath with needle-like spines, color bluish-green
FLOWERS: Small greenish 3-parted flowers on large spreading arrangement from crown
FRUITS: Nut-like drupes covered with short hairs
SEEDS: 1 per fruit
LIGHT: Partial shade
SOIL: Alluvium or sandy clay loam
MOISTURE: Wet to moist
SEASONAL ASPECT: None
ZONE: 1
USES: Specimen, for tropical effect
CARE: None
PROPAGATION BY: Seeds or seedlings
NATIVE OF: Coastal, S.C. to Miss.

Dwarf Palmetto

Sabal minor

Palm Family

GROWTH HABIT: Stout evergreen shrub from short underground stem
LEAVES: Fan-shaped, to 3′ broad, slightly bluish-green; petiole edges sharp but not spiny
FLOWERS: Small, greenish and in large clusters among the leaves
FRUITS: Globose, black, thin pulp, 1/3″ wide
SEEDS: 1 per fruit
LIGHT: Tolerant
SOIL: Silt loam
MOISTURE: Moist or wet
SEASONAL ASPECT: None
ZONE: 1, 2
USES: Border and background for low shrubs
CARE: None
PROPAGATION BY: Seedlings
NATIVE OF: Coastal s.e. United States
OTHER: A hardy plant of neglected ornamental value

Saw Palmetto

Serenoa repens

Palm Family

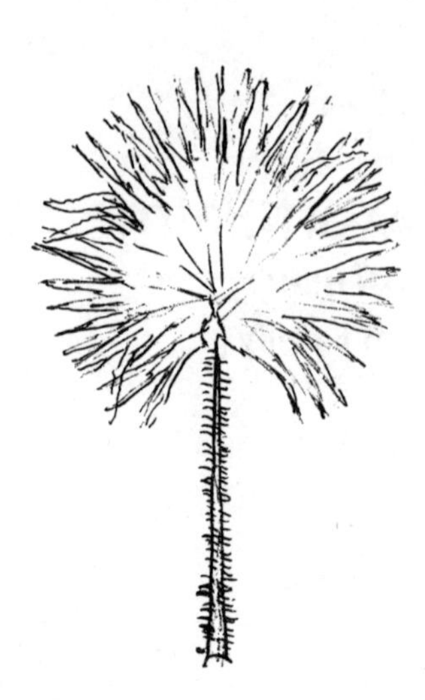

GROWTH HABIT: Low evergreen palm with trunk or rhizome commonly underground
LEAVES: Fan-shaped, yellow green, to 2′ wide
FLOWERS: A branching inflorescence mostly shorter than the leaves; flowers small and yellowish
FRUITS: Drupe-like and about 1″ long
SEEDS: One and grooved
LIGHT: Full sun
SOIL: Sand or sandy
MOISTURE: Moist to dry
SEASONAL ASPECT: None
ZONE: 1
USES: Interest or background
CARE: None
PROPAGATION BY: Seedlings or rhizomes
NATIVE OF: Coastal South Carolina and south
OTHER: Forms dense stand

Pampas Grass

Cortaderia selloana

Grass Family

GROWTH HABIT: Leafy, clump-forming perennial, dying back to ground level in fall
LEAVES: 3-6′ long arising from or near ground and abundantly produced
FLOWERS: Male and female on separate plants but in either case forming large silky panicles borne on culms much overtopping leaves
FRUITS: Borne in 2-several flowered spikelets
SEEDS: Very thin
LIGHT: Full sun
SOIL: Tolerant
MOISTURE: Moist
SEASONAL ASPECT: Panicles in summer and fall
ZONE: 1, 2, 3
USES: Interest, specimen or accent
CARE: Winter removal of dead top
PROPAGATION BY: Rhizomes
NATIVE OF: Chile
OTHER: One variety bears pink panicles

Partridge-Berry

Mitchella repens

Madder Family

GROWTH HABIT: Prostate creeping evergreen
LEAVES: Opposite, ovate, leathery, 1/2″ long
FLOWERS: White or tinted, paired
FRUITS: Scarlet berry
SEEDS: Few
LIGHT: Shade
SOIL: Most types if rich in humus
MOISTURE: Moist
SEASONAL ASPECT: Bright red berries in late summer and fall
ZONE: 1, 2, 3
USES: Ground cover
CARE: None
PROPAGATION BY: Seeds or rooted stems
NATIVE OF: e. United States
OTHER: Berries edible but not sweet; white fruited form known.

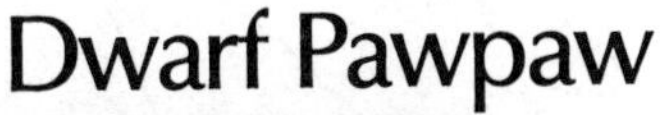

Dwarf Pawpaw

Asimina parviflora

Custard-Apple Family

GROWTH HABIT: Deciduous shrub to 8′ tall
LEAVES: Alternate, obovate, entire, smooth above, rusty, pubescent beneath, to 6″ long
FLOWERS: Before the leaves, brownish- or greenish-purple
FRUITS: 1″ long, berry-like, aromatic
SEEDS: 1-few flattened seeds
LIGHT: Partial sun
SOIL: Sandy or sandy loam
MOISTURE: Dry to moist
SEASONAL ASPECT: None
ZONE: 1, 2, 3
USES: Occasional shrub
CARE: None
PROPAGATION BY: Seedlings
NATIVE OF: s.e. United States
OTHER: *A. triloba* is larger, to 20′ tall, and adapted to low woods. Fruits somewhat edible.

Pear, Bradford or Stone

Pyrus calleryana var. bradfordii

Rose Family

GROWTH HABIT: Hardy, deciduous tree
LEAVES: Alternate, ovate
FLOWERS: White, slightly ill-scented, produced abundantly before the leaves
FRUITS: Hardy, green, 1/2″ wide
SEEDS: Few
LIGHT: Full sun
SOIL: Sandy loam with good fertility
MOISTURE: Moist to dry
SEASONAL ASPECT: Spring flowers which precede the leaves
ZONE: 1, 2, 3
USES: Street, freestanding
CARE: None
PROPAGATION BY: Cuttings or root sprouts
NATIVE OF: China
OTHER: *P. communis*; Europe to w. Asia; pear, cultivated in many varieties.

Pearl-Bush

Exochorda racemosa

Rose Family

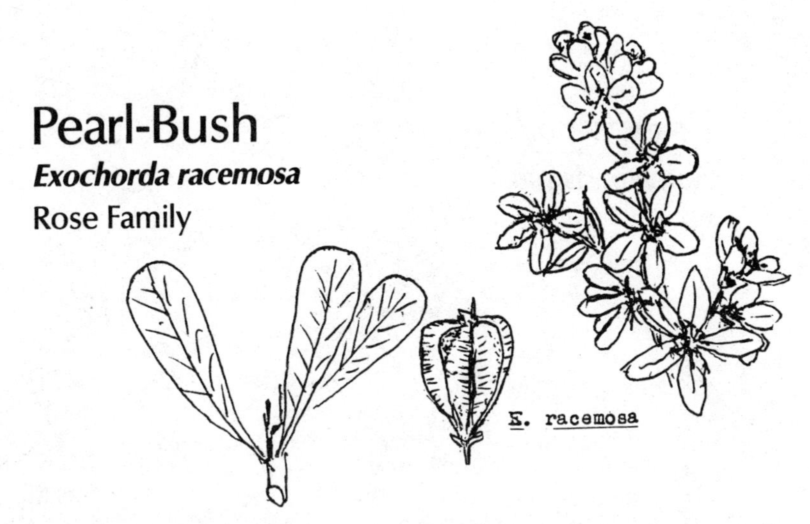

GROWTH HABIT: Slender, deciduous, much-branched shrub to 10′ high or more

LEAVES: Alternate, broadly elliptic, entire on flowering branches, toothed elsewhere, to 2″ long

FLOWERS: White, 1 1/2″ across and in short racemes

FRUITS: 1/2″ long, dry and narrowly winged

SEEDS: Several

LIGHT: Partial shade

SOIL: Fertile

MOISTURE: Moist

SEASONAL ASPECT: Flowers and new leaves in early spring

ZONE: 1, 2, 3

USES: Occasional flowering shrub

CARE: None

PROPAGATION BY: Seeds or cuttings

NATIVE OF: China

OTHER: *E. giraldii wilsonii*; China; somewhat larger, more vigorous and more floriferous.

Pepper-Vine

Ampelopsis spp.

Grape Family

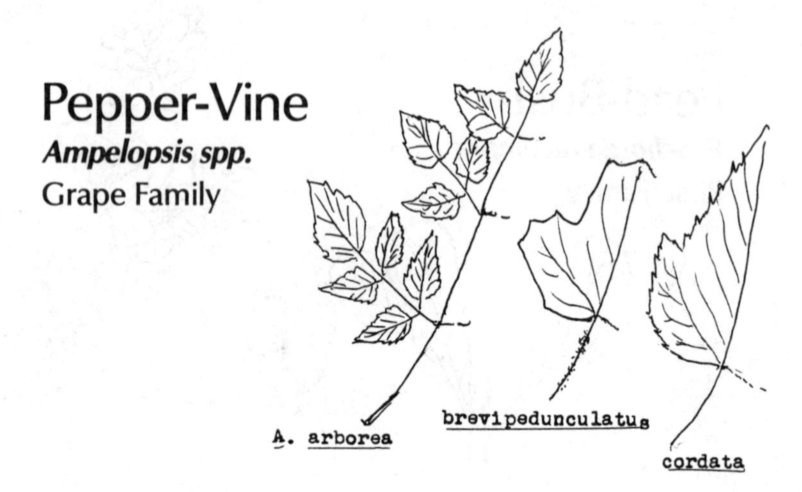

GROWTH HABIT: Deciduous, tendril-bearing, woody vines, tendrils forking but not dilated at tips

LEAVES: Alternate, simple, cordate, toothed and 2-4″ wide, or bipinnate with leaflets to 1″ long and coarsely toothed

FLOWERS: Perfect or unisexual, small and mostly 5-parted

FRUITS: Bluish to black grape-like berries to 1/3″ wide

SEEDS: 1-3

LIGHT: Partial shade

SOIL: Alluvial

MOISTURE: Moist to wet

SEASONAL ASPECT: None

ZONE: 1, 2

USES: Tree, wall, trellis

CARE: None

PROPAGATION BY: Seedlings

NATIVE OF: See below

OTHER: *A. brevipedunculata*; Japan, Manchuria, China; leaves cordate, petioles of young leaves and twigs hairy.

A. cordata; s.e. United States; leaves cordate, twigs and petioles of young leaves usually glabrous; trunk becoming 5″ in diameter toward base and developing dark, deeply checkered bark; a high climber.

A. arborea; s.e. United States; leaves bipinnate; stem to 2″ in diameter and conspicuously jointed every 4-8″.

Periwinkle, Running Myrtle

Vinca minor

Dogbane Family

GROWTH HABIT: Trailing evergreen subshrub forming a dense cover to 5″ deep; several varieties are cultivated
LEAVES: Opposite, leathery, elliptic, glossy with light colored midvein and to 1 1/2″ long
FLOWERS: Blue or bluish, 1/2″ wide or wider with open center
FRUITS: Slender 3″ long pods
SEEDS: Small
LIGHT: Partial shade
SOIL: Any type with good top soil layer
MOISTURE: Moist
SEASONAL ASPECT: None
ZONE: (1), 2, 3
USES: Ground cover, rock garden, bank
CARE: None
PROPAGATION BY: Rooted runners
NATIVE OF: Europe
OTHER: *V. major*; Europe; evergreen subshrub with weakly ascending stems; ovate leaves to 2″ long, larger blue flowers and forming a rather thin ground cover about 1′ deep. Forms with leaves variegated with white or yellow margins or spots are known.

Pernettya

Pernettya mucronata

Heath Family

GROWTH HABIT: Much branched, evergreen shrub to 2′ high
LEAVES: Ovate, 1/2″ long, toothed and sharp-pointed
FLOWERS: 1/4″ long, white or pinkish, perfect but 2 strains needed for lots of fruit
FRUITS: 1/2″ wide, white to red or purple berry
SEEDS: Several
LIGHT: Full sun, and it will remain a neat shrub
SOIL: Any good type
MOISTURE: Moist
SEASONAL ASPECT: Late spring flowers; fall and winter fruits
ZONE: 1, 2, 3
USES: Specimen
CARE: None, if grown in full sun
PROPAGATION BY: Seedlings or cuttings
NATIVE OF: Chile

Persimmon

Diospyros virginiana
Ebony Family

GROWTH HABIT: Deciduous tree, usually small, with dark gray checkered bark and hard white wood
LEAVES: Alternate, ovate to elliptic, to 5″ long, entire with crescent-shaped vascular trace where leaf separates from twig
FLOWERS: Cream or greenish, some male only, almost 1/2″ long, fragrant
FRUITS: Orange 1 - 1 1/2″ wide and quite edible after frost, calyx much enlarged and woody
SEEDS: Several, each almost 1/2″ long
LIGHT: Lots of sun
SOIL: Most any
MOISTURE: Dry-moist
SEASONAL ASPECT: Flowers soon after leaves, autumn foliage is multicolored
ZONE: 1, 2, 3
USES: Freestanding
CARE: None
PROPAGATION BY: Seeds or seedlings
NATIVE OF: e. United States
OTHER: A valuable wood, long used by the textile industry for picker-sticks and shuttles. Also used for heads of golf clubs.
Japanese Persimmon, *D. kaki*, has leaves to 7″ long and fruit to 3″ wide.

Phillyrea

Phillyrea decora

Olive Family

GROWTH HABIT: Evergreen shrub to 10′ high
LEAVES: Opposite, entire, long-elliptic and to 5″ long
FLOWERS: Small, white, axillary, few together and 4-parted
FRUITS: Black, fleshy and 1/2″ wide
SEEDS: 1
LIGHT: Sun or partial shade
SOIL: Tolerant to most
MOISTURE: Dry
SEASONAL ASPECT: None
ZONE: 1, 2, 3
USES: Hedge, screen, interest
CARE: Prune for shape
PROPAGATION BY: Seeds and cuttings
NATIVE OF: Black Sea region

Phlox, Moss-Pink

Phlox subulata

Phlox Family

GROWTH HABIT: Semievergreen, prostrate, subshrub forming dense mat to 4″ thick
LEAVES: Opposite or crowded at nodes, very narrow and long pointed, to 1/2″ long, or longer on vigorous shoots
FLOWERS: Many, purplish or pink and raised just above the mat, 5-parted
FRUITS: 3-valved capsule
SEEDS: Tiny
LIGHT: Full sun
SOIL: Clay loam
MOISTURE: Dry or moist
SEASONAL ASPECT: Early spring flowers
ZONE: (1), 2, 3
USES: Ground covei, bank, rockery
CARE: None
PROPAGATION BY: Rooted stems
NATIVE OF: e. United States
OTHER: *P. nivalis*; e. United States; similar except the stamens are hidden within the corolla tube and the corolla lobes have almost entire margins.

Photinia, Red-Tip

Photinia spp.

Rose Family

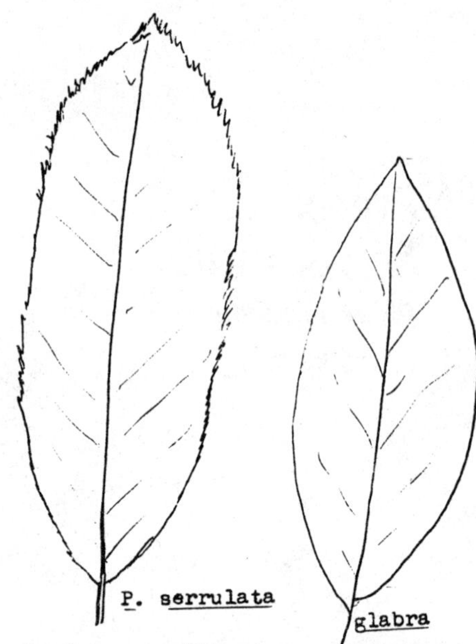

GROWTH HABIT: Evergreen shrubs or small trees
LEAVES: Alternate, finely toothed, 2-8″ long, red when young
FLOWERS: Small, white, and in large, rather flat-topped clusters
FRUITS: Red, berry-like and 1/4″ wide
SEEDS: 1-4
LIGHT: Lots of sun
SOIL: Fertile
MOISTURE: Moist
SEASONAL ASPECT: New growth is red
ZONE: 1, 2, 3
USES: Screen, hedge, occasional shrub, background
CARE: Maintaining desired shape and size
PROPAGATION BY: Cuttings
NATIVE OF: s. and e. Asia
OTHER: *P. glabra*; Red-Tip; Japan; smaller shrub with smaller leaves, 2-4″ long, that are conspicously red when young.

P. serrulata; Large Photinia; China; large shrub with sharply-toothed leaves 4-8″ long.

Pinckneya

Pinckneya pubens

Madder Family

GROWTH HABIT: Deciduous shrub or small tree
LEAVES: Opposite, elliptic or broader, entire
FLOWERS: About 1″ long, cream with maroon spots; sepals 5, one often much enlarged and white or pink
FRUITS: Small capsule
SEEDS: Several
LIGHT: Partial shade
SOIL: Silty or clay loam
MOISTURE: Moist to wet
SEASONAL ASPECT: Summer flowers
ZONE: 1, (2)
USES: Specimen
CARE: None
PROPAGATION BY: Seedlings or cuttings
NATIVE OF: s.e. United States coastal plain

Pine

Pinus spp.

Pine Family

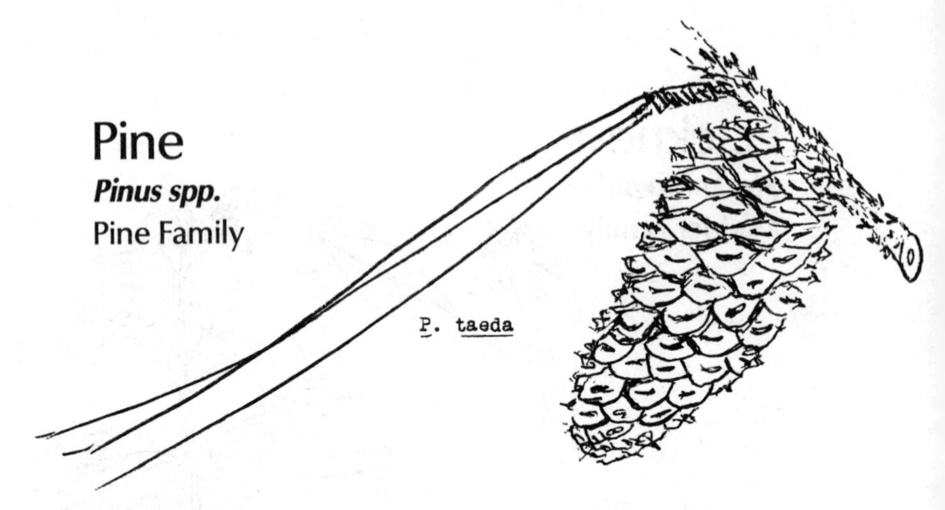

GROWTH HABIT: Evergreen trees with one main trunk
LEAVES: Needle shaped and in bundles of 2, 3 or 5
FLOWERS: Male as yellowish catkins in early spring; female as woody cones in summer
FRUITS: Woody cone
SEEDS: Winged nutlets
LIGHT: Full sun
SOIL: Tolerant of most types
MOISTURE: Moist or dry
SEASONAL ASPECT: None
ZONE: See below
USES: For high canopy, open, thin shade, framing or freestanding
CARE: None
PROPAGATION BY: Seeds or seedlings
NATIVE OF: eastern United States or as noted below
OTHER: For Zones 1 and 2:

P. australis; Longleaf; s.e. United States; leaves to 15″ long and in clusters of 3, crowded toward branch tips; cones to 10″ long.

P. taeda; Loblolly; s.e. United States; leaves to 6″ long, in clusters of 3; cones to 5″ long.

P. echinata; Yellow or Shortleaf; e. United States; leaves to 3 1/2″ long, in clusters of 2; cones to 2 1/2″ long

For Zones 2 and 3:

P. virginiana; Virginia, e. United States; leaves to 2 1/2″ long, twisted, in clusters of 2; cones to 2 1/2″ long.

P. strobus; White, e. United States; leaves to 7″ long; cones to 8″ long; branches whorled.

P. sylvestris; Scotch; Europe, Asia; leaves to 2 1/2″ long; cones to 2″ long, leaves in 2′s.

P. mugo; Alps; dwarf variety grown as bushy shrub, to 5′ high, tree form exists, leaves in 2′s.

Pineapple Guava

Feijoa sellowiana

Myrtle Family

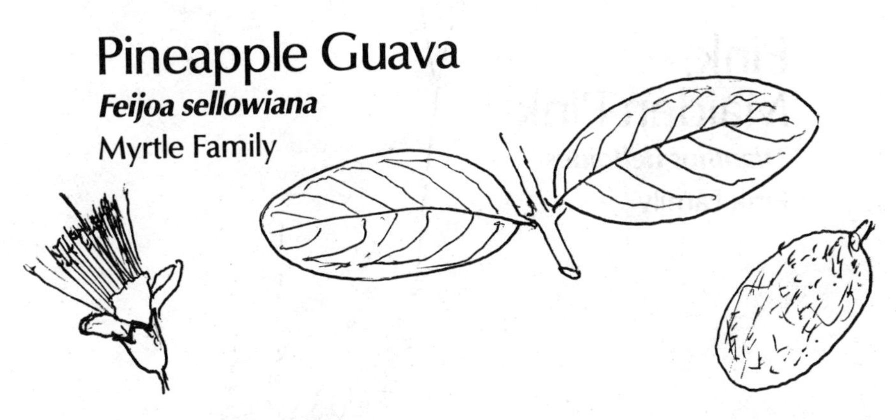

GROWTH HABIT: Semievergreen shrub grown for ornament and fruit

LEAVES: Opposite, green above, silvery pubescent beneath, elliptic and to 2 1/2″ long

FLOWERS: 1″ or more wide, white hairy outside, purplish within, stamens long and red

FRUITS: Oval, to 2 1/2″ long, dull green to yellowish or reddish and pulpy inside; somewhat edible

SEEDS: Several

LIGHT: Full or partial sun

SOIL: Fertile

MOISTURE: Moist

SEASONAL ASPECT: Summer flowers

ZONE: 1, 2

USES: Occasional shrub, specimen, interest, fruit

CARE: None

PROPAGATION BY: Seedlings

NATIVE OF: s. America

Pink, Maiden Pink

Dianthus deltoides

Pink Family

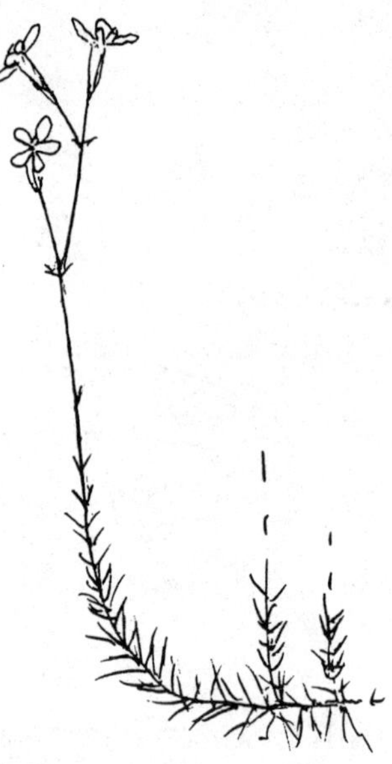

GROWTH HABIT: Low tufted evergreen forming loose mat
LEAVES: Opposite, narrow, sharp-pointed and powdery green
FLOWERS: White, 1/2′ wide and raised 1′ high on forking stems
FRUITS: Half inch long capsules
SEEDS: Numerous
LIGHT: Plenty of Sun
SOIL: Most any type
MOISTURE: Dry to moist
SEASONAL ASPECT: Summer flowers
ZONE: 2, 3
USES: Rock garden
CARE: Occasional weeding
PROPAGATION BY: Seeds
NATIVE OF: Great Britain, Japan
OTHER: Many varieties.
Chinese Pink, *D. chinensis*; plant taller than above; several varieties.
Cottage Pink, *D. plumarius*, w. Europe, Asia; petals with fringed margins; several varieties.

Pistacia

Pistacia chinensis

Cashew Family

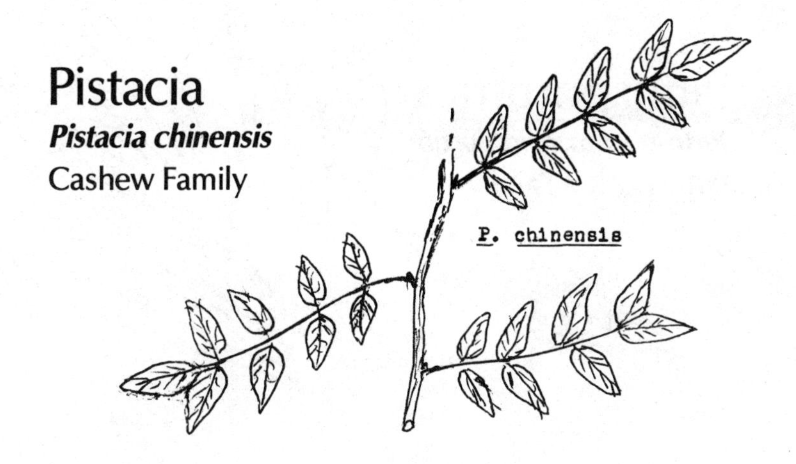

GROWTH HABIT: Medium sized, deciduous tree

LEAVES: Alternate, pinnate, 10-12 lanceolate, entire and unequal-sided leaflets, reddening in autumn; leaflets 1 1/2-3″ long

FLOWERS: Male and female on different plants, both small, without petals and in large panicles

FRUITS: Very small, dry drupes, less than 1/2″ wide, red at maturity

SEEDS: 1

LIGHT: Full sun

SOIL: Most any fertile, loose type

MOISTURE: Moist

SEASONAL ASPECT: Red or crimson leaves in fall

ZONE: 1, 2

USES: Freestanding or specimen

CARE: None

PROPAGATION BY: Cuttings or seedlings

NATIVE OF: China

OTHER: *P. vera*; Mediterranean region; produces the pistachio nut of commerce. May be grafted to above.

Pittosporum

Pittosporum viridiflorum

Pittosporum Family

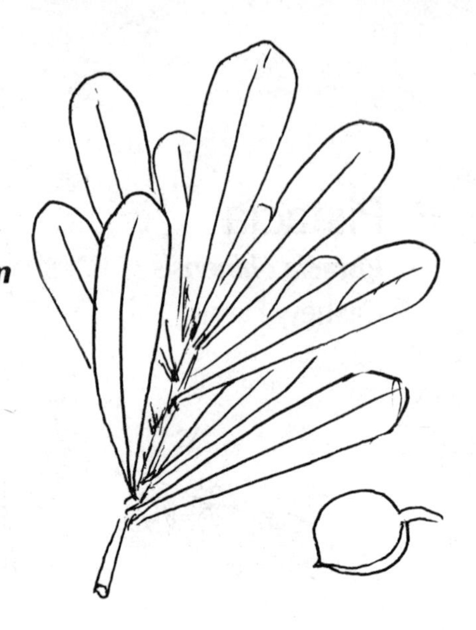

GROWTH HABIT: Much-branched evergreen shrub to 20′
LEAVES: Smooth, thick, entire, obovate with rounded tips, 2-3′ long, alternate but appearing clustered toward the tips of young growth
FLOWERS: Greenish-yellow to whitish, 1/2′ wide, fragrant and in small clusters
FRUITS: A roundish, leathery, 2-4-valved capsule, 1/2″ wide
SEEDS: 2
LIGHT: Full or partial sun
SOIL: Any good loam
MOISTURE: Moist
SEASONAL ASPECT: None
ZONE: 1, 2, (3)
USES: Shrubbery, shrub border, unit arrangement, hedge, occasional shrub
CARE: Some clipping
PROPAGATION BY: Seedlings
NATIVE OF: s. Africa
OTHER: A variegated form of the above and a few other species exist. *P. tobira*; Japan; an excellent choice for zone 1.

Plum Yew

Cephalotaxus spp.

Plum-Yew Family

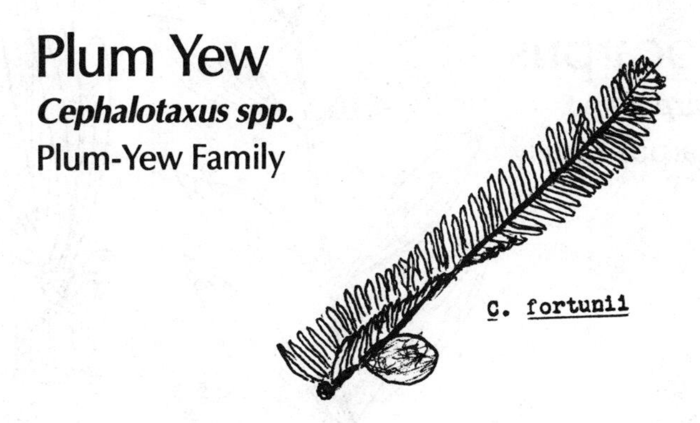

GROWTH HABIT: Evergreen shrubs or small trees similar in appearance to Taxus
LEAVES: Linear, somewhat 2-ranked and 1-3″ long
FLOWERS: Male and female flowers usually on different plants; male as globose heads about 1/2″ long; female as several bracts becoming 1-2 seeded fleshy drupe
FRUITS: 1″ long drupe-like and purplish
SEEDS: 1-2 per fruit
LIGHT: Full or partial sun
SOIL: Fertile loam
MOISTURE: Moist
SEASONAL ASPECT: None
ZONE: 1, 2
USES: Interest, informal shrub border
CARE: None
PROPAGATION BY: Seedlings
NATIVE OF: China and Japan
OTHER: *C. fortunii*, Chinese Plum Yew, leaves gradually pointed and 1-2″ long.
C. harringtonia, Japanese Plum Yew, leaves abruptly pointed and 2-3″ long.

Podocarpus

Podocarpus spp.

Podocarpus Family

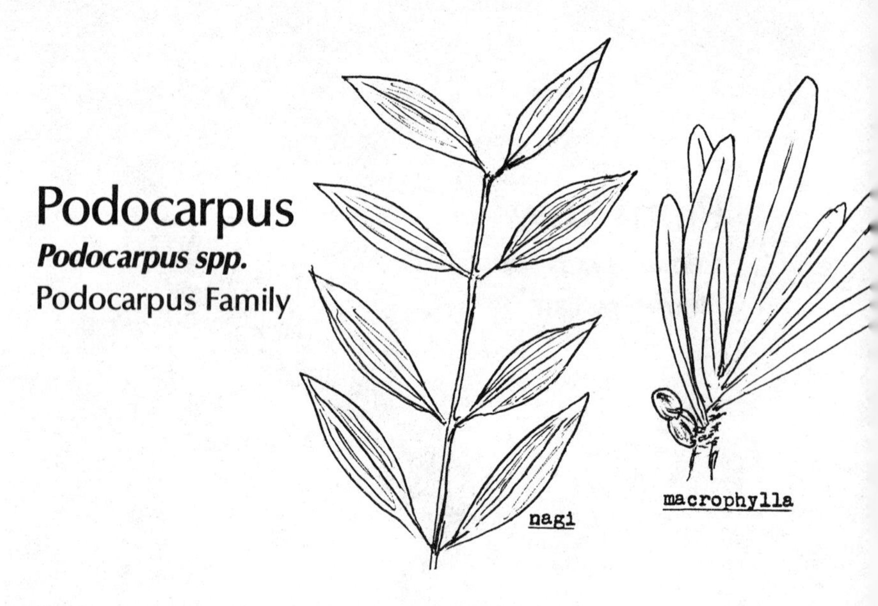

GROWTH HABIT: Evergreen shrubs with drooping branches; may attain small tree size

LEAVES: Alternate or opposite, entire, dark green or lusterous

FLOWERS: Staminate in 1-1 1/2″ long catkins; female as ovule surrounded by scale

FRUITS: Purplish, pulp covered, drupe like, 1/2″ wide and on fleshy stalk

SEEDS: 1

LIGHT: Shade or sun

SOIL: Tolerant of many types

MOISTURE: Moist

SEASONAL ASPECT: None

ZONE: 1, 2, 3

USES: Accent, espalier, shrub border

CARE: Pruning for desired size and shape

PROPAGATION BY: Seedlings

NATIVE OF: Japan

OTHER: *P. macrophylla*, leaves clustered toward twig tips and less than 1/2″ wide; dark green above.

P. nagi, leaves opposite, from 1/2-1″ wide and bright shining green above. Excellent for cut greenery.

Pomegranate

Punica granatum

Pomegranate Family

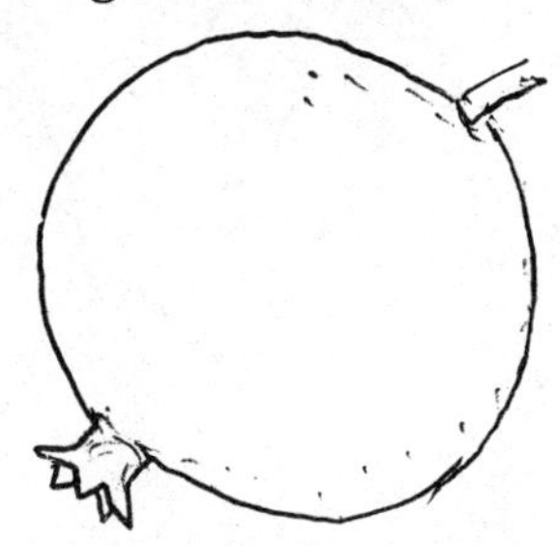

GROWTH HABIT: Large, much-branched deciduous shrub grown for ornament and fruit
LEAVES: Oblong to lanceolate, to 2″ long and glabrous
FLOWERS: Orange-red and 1 1/2″ long, calyx thick and also orange-red
FRUITS: About the size of an orange, reddish in color and with juicy pulp surrounding each seed
SEEDS: Many
LIGHT: Full sun for best development
SOIL: Fertile
MOISTURE: Moist
SEASONAL ASPECT: None
ZONE: 1, 2
USES: Occasional flowering shrub, specimen, fruit
CARE: Little
PROPAGATION BY: Cuttings and seedlings
NATIVE OF: s. Asia

Popcorn-Tree, Chinese Tallow-Tree

Sapium sebiferum

Spurge Family

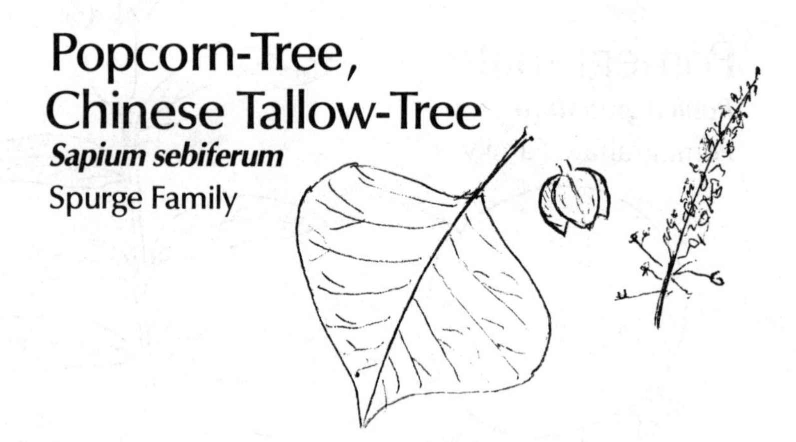

GROWTH HABIT: Small to medium deciduous tree with milky juice
LEAVES: Alternate, very broadly ovate, smooth, entire, pointed, to 3″ long and reddening in age
FLOWERS: Male and female separate but on same tree, without petals and borne in terminal or lateral spikes or clusters
FRUITS: Leathery 3-lobed capsule 1/2″ wide
SEEDS: 3 and white
LIGHT: Nearly full sun
SOIL: Tolerant
MOISTURE: Moist
SEASONAL ASPECT: Fall when capsules open and expose white seeds
ZONE: 1, 2
USES: Freestanding or specimen
CARE: None
PROPAGATION BY: Seedlings
NATIVE OF: China, Japan
OTHER: The waxy acid covering has been used for candles, soap and dressing for fabrics; juice and seeds toxic, foliage may produce a rash.

Poplar, Cottonwood

Populus spp.

Willow Family

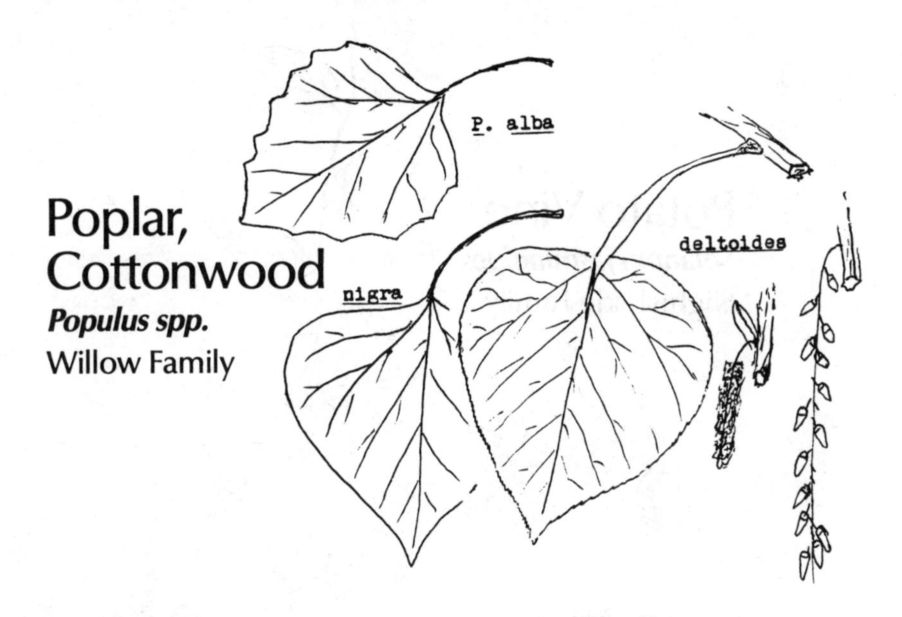

GROWTH HABIT: Large deciduous trees with spreading or narrow crowns
LEAVES: Alternate, broad, most with petioles flattened vertically
FLOWERS: Male and female borne separate, but both in catkins
FRUITS: Small capsule
SEEDS: Small and with much white down
LIGHT: Full or partial sun
SOIL: Loam or silt loam
MOISTURE: Moist or wet
SEASONAL ASPECT: None
ZONE: 1, 2, 3
USES: Freestanding, border tree, along wet bank, windbreak
CARE: None
PROPAGATION BY: Seedlings and suckers
NATIVE OF: e. North America
OTHER: White Poplar, *P. alba*; bark and underside of leaves white; suckers badly.

Lombardy Poplar, *P. nigra var. italica*; tall with very narrow crown; suckers badly.

Southern Cottonwood, *P. deltoides*; large tree for wet places.

Swamp Cottonwood, *P. heterophylla*, similar to above but with pubescent leaves, mainly beneath and when young.

Potato Vine

Solanum jasminoides

Nightshade Family

GROWTH HABIT: Weakly twining semievergreen vine
LEAVES: Upper 1-3″ long single and entire, lower often with small lateral lobes
FLOWERS: Star-shaped, 1″ wide and white with bluish tinge
FRUITS: 1/2″ wide yellowish berry
SEEDS: Several
LIGHT: Full sun
SOIL: Sandy loam
MOISTURE: Moist
SEASONAL ASPECT: Flowers in summer
ZONE: 1
USES: Fence, shed, trellis
CARE: Prune to prevent a tangled mass
PROPAGATION BY: Seeds
NATIVE OF: Brazil
OTHER: If killed back by severe cold will usually come back in spring.

Prickly Pear

Opuntia vulgaris

Cactus Family

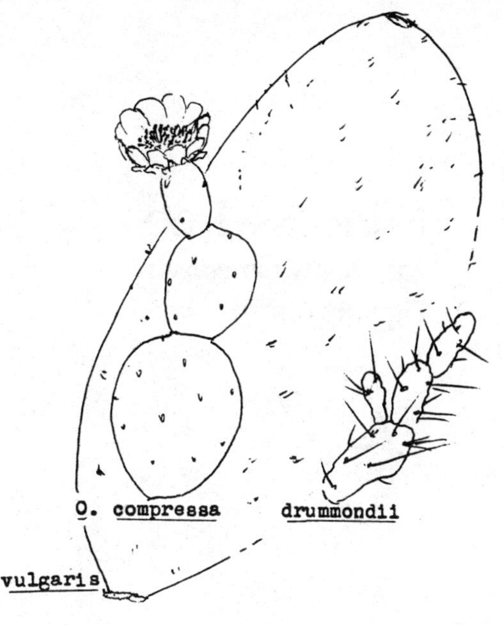

GROWTH HABIT: To 3′ high; joints flattened and from 8-16″ long; evergreen
LEAVES: Alternate, fleshy; 1/2″ long and promptly lost
FLOWERS: Yellow and to 3″ broad
FRUITS: Purple to reddish-brown berry to 2 1/2″ long
SEEDS: Numerous and dark
LIGHT: Full sun
SOIL: Sandy loam
MOISTURE: Dry to moist
SEASONAL ASPECT: Spring flowers
ZONE: 1, 2, 3
USES: Interest, specimen
CARE: None
PROPAGATION BY: Seeds and cuttings
NATIVE OF: Brazil and Argentina
OTHER: *O. compressa*; e. United States, much smaller than above, otherwise similar; joints 2-8″ long, not easily detachable; spines short.
O. drummondii; e. United States; segments 1/2-2″ long; very easily detachable; spines to 2″ long. The short tender leaves usually present in early summer are soon lost but the flattened, green, jointed stem carries on their function.

Princess-Tree

Paulownia tomentosa

Figwort Family

GROWTH HABIT: Deciduous tree resembling Catalpa except for the seed pods
LEAVES: Opposite, heart-shaped, entire or 3-lobed, to 10″ long, downy beneath
FLOWERS: Pale violet with darker spots, 2″ long and showy in 1′ long terminal panicles
FRUITS: 1 1/2″ long woody capsules
SEEDS: Many and winged
LIGHT: Partial or full sun
SOIL: Very tolerant
MOISTURE: Moist
SEASONAL ASPECT: Flowers before leaves appear
ZONE: 1, 2, 3
USES: Background, specimen
CARE: It spreads and has become naturalized in many areas from New York south.
PROPAGATION BY: Seedlings
NATIVE OF: China

Cherry Prinsepia

Prinsepia spp.

Rose Family

GROWTH HABIT: Deciduous shrub, usually spiny; to 6′
LEAVES: Alternate, ovate to lanceolate, toothed and appearing very early
FLOWERS: Yellow, 1/2″ wide and 1-3 in leaf axils
FRUITS: 1/2″ wide, red or purple, late summer, edible
SEEDS: 1
LIGHT: Full sun or light shade
SOIL: Any fertile type
MOISTURE: Moist
SEASONAL ASPECT: Early spring flowers
ZONE: 2, 3
USES: Hedge or barrier
CARE: Prune only to remove overcrowded branches
PROPAGATION BY: Cuttings
NATIVE OF: Manchuria
OTHER: *P. sinensis*; n. China; leaves oblong, with marginal hairs and few teeth; flowers yellow; fruits purple and 1/2″ wide.

P. mollis; Himalayas; to 12′ high; leaves elliptic and toothed; flowers white and fragrant.

Common Privet

Ligustrum vulgare

Olive Family

GROWTH HABIT: Commonly a deciduous shrub widely grown in several varieties exhibiting various habits
LEAVES: Opposite, ovate to lanceolate, entire, glabrous and to 2″ long
FLOWERS: Small, white, numerous, fragrant and in panicles
FRUITS: Small black berries
SEEDS: 1
LIGHT: Sun or shade
SOIL: Almost any
MOISTURE: Wet, moist or dry
SEASONAL ASPECT: Flowers soon after leaves are grown
ZONE: 1, 2, 3
USES: Hedge
CARE: Trim for desired size and shape
PROPAGATION BY: Cuttings and seedlings
NATIVE OF: China
OTHER: *L. sinense*; China; confusingly similar but instead of 2″ long panicles, as in *L. vulgare*, may bear 4″ long panicles. Both are extremly hardy although *vulgare* in some areas is subject to a deadly blight.

Golden Privet; *L. X vicary*; a hardy evergreen small-leaved hybrid between a gold variety of *L. ovalifolium* and *L. vulgare.*

Quince

Cydonia oblonga

Rose Family

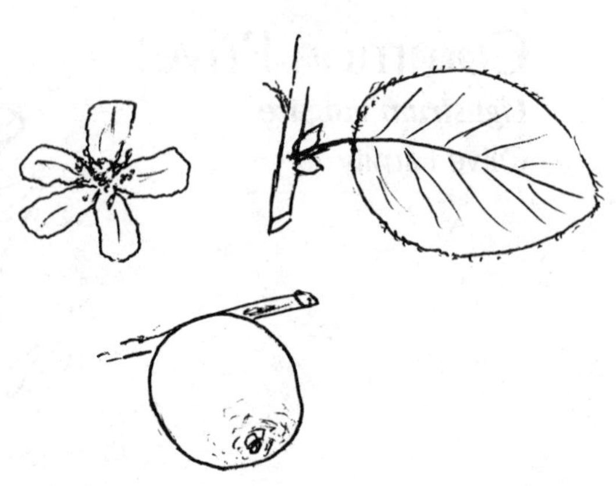

GROWTH HABIT: Small, deciduous, round-headed fruit-bearing tree
LEAVES: Alternate, oblong to broadly so, short-petioled, smooth above at maturity, pubescent beneath and entire
FLOWERS: White to pink, 2″ wide and appearing with the leaves
FRUITS: Yellow, fragrant, fuzzy and to 3″ wide
SEEDS: Several
LIGHT: Full sun
SOIL: Tolerant
MOISTURE: Dry to moist
SEASONAL ASPECT: Flowering as the leaves appear
ZONE: 1, 2, 3
USES: Occasional shrub, interest, fruit
CARE: None
PROPAGATION BY: Seedlings
NATIVE OF: Iran

Redbud, Judas-Tree

Cercis canadensis

Legume Family

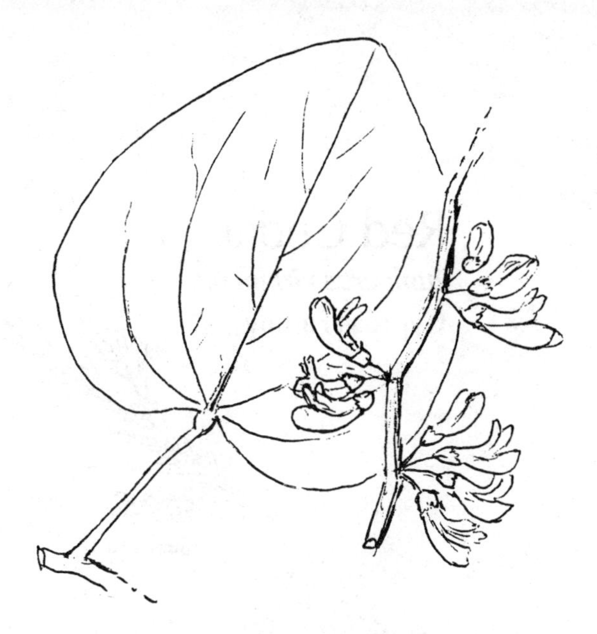

GROWTH HABIT: Spreading deciduous tree of small to medium size
LEAVES: Alternate, somewhat heart-shaped but broader than long; entire
FLOWERS: Rosey-pink, 1/2″ long in small clusters along twigs before leaves
FRUITS: Flat brown pod 2 - 3″ long
SEEDS: 3-6
LIGHT: Partial shade
SOIL: Tolerant
MOISTURE: Moist
SEASONAL ASPECT: Early spring flowers
ZONE: 1, 2, 3
USES: Freestanding, background, interest
CARE: None
PROPAGATION BY: Seedlings
NATIVE OF: e. United States
OTHER: The fresh flowers are sometimes added to a tossed salad or rolled in light batter and fried.

Red Cedar

Juniperus virginiana

Cypress Family

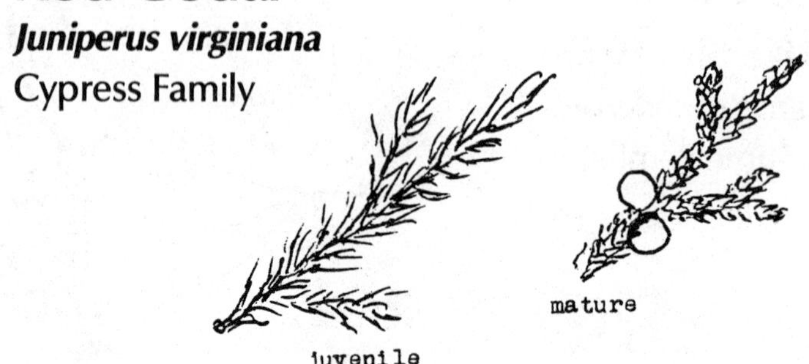

GROWTH HABIT: Common, hardy, mildly aromatic evergreen tree long in wide use as an ornament. Many forms are in cultivation as columnar, pyramidal, with pendulous branches, and with green, bluish or yellowish foliage.

LEAVES: Opposite, needle-like on juvenile plants and scale-like on mature

FLOWERS: Male as minute yellow catkins; female as minute greenish structures becoming berries

FRUITS: Blue berry-like cones, succulent and aromatic

SEEDS: 1 or 2 and wingless

LIGHT: Full sun

SOIL: Does well on most, even very infertile soils

MOISTURE: Dry

SEASONAL ASPECT: None

ZONE: 1, 2, 3

USES: Freestanding, perimeter, roadside, framing

CARE: None

PROPAGATION BY: Seedlings

NATIVE OF: e. United States

OTHER: In addition to the tree-like kinds a number of shrubby kinds have been developed.

Redwood

Sequoia sempervirens

Taxodium Family

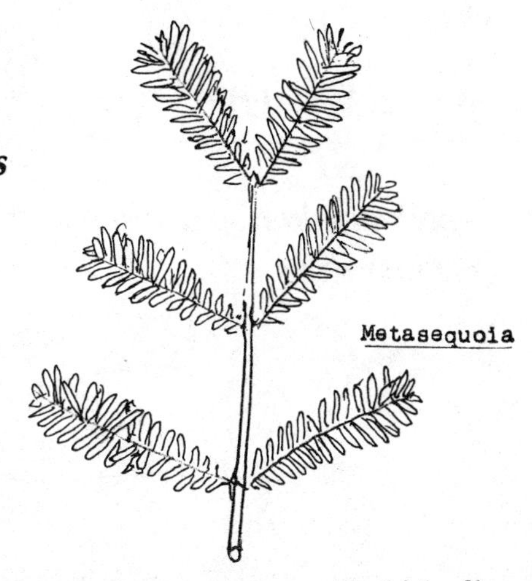

Very large and tall evergreen from the west coast with thick, fibrous, reddish bark, very small alternate leaves of two kinds, linear and scale like, woody cones about 1″ long; and winged seeds. Commonly propagated by sprouts from root burls.

Sequoiadendron giganteum, Giant Redwood, California. *Metasequoia glyptostroboides*, Dawn Redwood, China. Both are occasionally planted in the southeast and like the Redwood above may be regarded as specimen or interest trees.

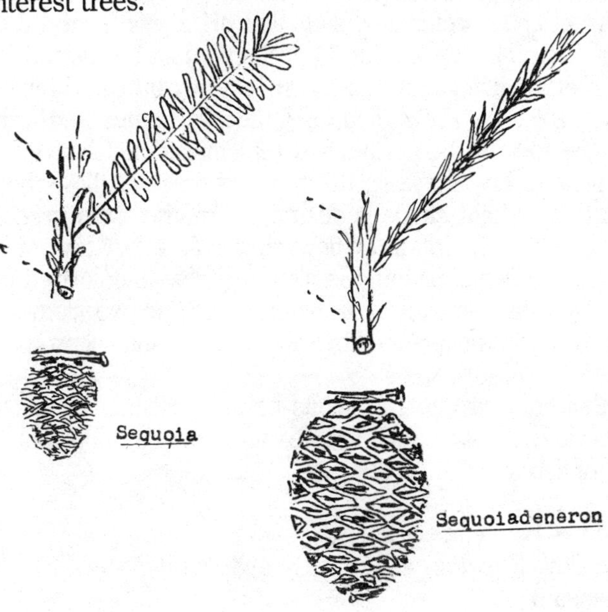

Rhododendron, Azalea

Rhododendron spp. and XX

Heath Family

The genus *Rhododendron*, including plants popularly referred to as Rhododendrons (evergreen) and Azaleas (deciduous), is one of the most commonly cultivated and best known groups of ornamentals. Almost 700 species are recognized. When varieties and hybrids are added, the list approaches 2500 and makes this the largest group of ornamental plants.

Thriving best in cool temperatures, they are widely distributed over the cool and temperate regions of earth, particularly in the northern hemisphere. They prefer acid soils and pH levels from 4.5 to 6.5 are generally favorable. Fertilizing is perhaps most easily and adequately accomplished by using some of the specifically prepared mixes. Mulching is desirable. As the mulch layer decays humus and minerals are added to the soil and the periodically replenished mulch layer serves the dual purpose of helping keep the roots of these shallow-rooted plants cool. Shifting or light shade is desirable. Another step in maintenance is pruning. Twig-clipping after flowering is an effort toward shape and size control. Renewal pruning is also desirable and carried out once a year involves the removal of an older branch or two from some point near the base of the plant. This allows new and vigorous shoots to develop.

With this huge group, the scope of this presentation allows only a brief overview. A few native species and several important classes of hybrids are listed.

Evergreen — native species

Mt. Rose Bay; *R. catawbiense*; to 15′ high, with flowers 2″ wide; leaves pale beneath.

Great Laurel; *R. maximum*; to 25′ high with greenish-white flowers to 1 1/2″ wide and leaves to 10″ long.

Carolina Rhododendron, *R. minus* (*R. carolinianum*); to 8′ high with rose or purplish flowers to 1″ wide; leaves to 3″ long and brownish beneath.

Evergreen — classes of hybrids

Catawba Hybrids; oldest and most common, hardy and dependable.

Maximum Hybrids; flowers tend to be hidden by the rapidly developing large leaves.

Fortune Hybrids; developed from the Chinese (*R. fortunei*); these produces flowers to 4″ wide.

Decidous — native species

Flame Azalea; *R. caldendulaceum*; to 8′ high with orange-yellow flowers to 2″ wide.

Wild Honeysuckle, *R. canescens*; to 12′ high with pink or white flowers and a soft-hairy carolla tube.

Pinxter Flower, *R. nudiflorum*; to 8′ high with white or pink flowers to 1 1/2″ wide.

Dwarf Azalea, *R. atlanticum*; to 3′ high, stoloniferous; flowers white or pink.

Swamp Honeysuckle, *R. viscosum*, similar to *R. nudiflorum*.

Deciduous — classes of hybrids

Kurume Hybrids; over 200 have been developed from the Chinse *R. obtusum*; 50 or more are available in the United States; flowers white to red and to 1 1/2″ wide. Many are often used as pot plants.

Indian Hybrids; flowers white to rose and to 3″ across; not very cold hardy.

Ghent Hybrids; flowers from yellow to white; cold hardy.

Knap Hill Hybrids; includes the Exbury and New Zealand groups.

Flame Azalea

Roses

Rosa spp.

Rose Family

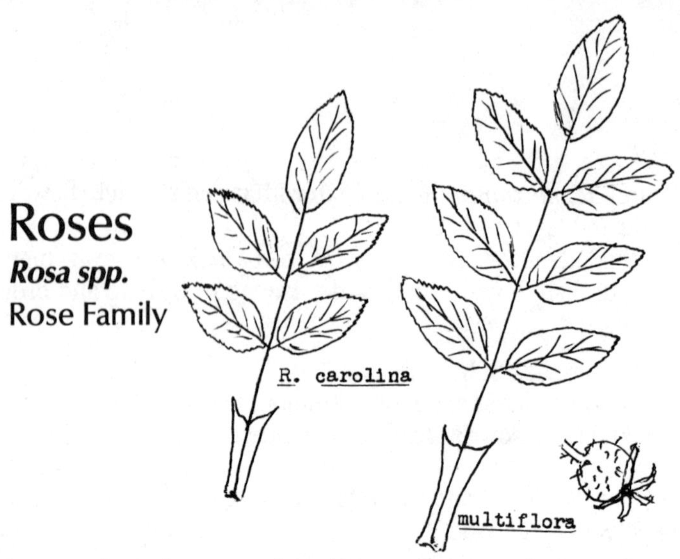

This genus includes more than 100 species, mostly from the temperate regions of the northern hemisphere. Because of their great popularity and ornamental significance, indoors and out, perhaps as many as a few thousand varieties, forms, clones and hybrids have been named. Some cultivated types whose popularity now may have declined, have long persisted around abandoned homesites and occasionally represent pleasant discoveries today. In cases where the original was a grafted specimen it may have lost its scion, or grafted on above ground part and remained as only the sprout development from the underground part.

The following species represents a few of the important ones.

NativeSpecies

Carolina Rose; *R. carolina*; Wild Rose. Slender, deciduous shrub to 3′; prickles mostly straight and weak; bristle-like hairs interspersed. Includes *R. virginiana.*

Swamp Rose; *R. palustris*. Deciduous shrub to 6′, with 5-9 leaflets, broad-based curved thorns and flowers usually 2 or more together (includes *R. floridana* and *R. virginiana*).

MultifloraSpecies

R. multiflora; Japan. Vigorous, somewhat arching shrub with flattened, curved thorns, leaflets mostly seven; flowers many together, white and fragrant.

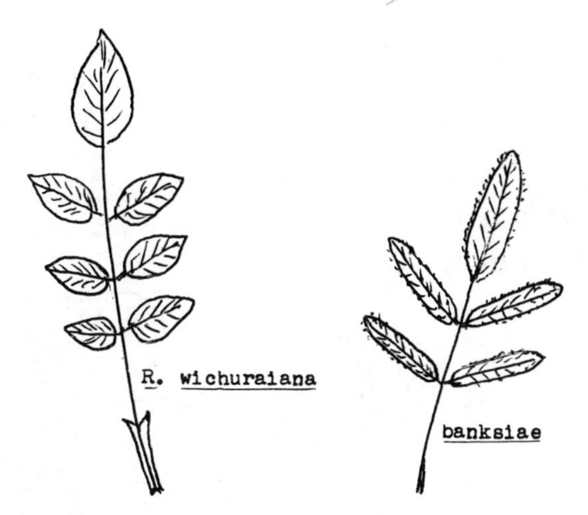

R. cathayensis; China. Similar to above but with somewhat larger, pink and less fragmrant flowers in clusters of only 10-20. Because of its hardiness and color it has been crossed with many other species and hybrids.

Ground Cover Species

Memorial Rose; *R. wichuraiana*; Japan. Evergreen trailing or leaning shrub with shining foliage and few to several white flowers per cluster. The Dorothy Perkins was developed from this species.

Tea Rose Species

R. odorata; China. Weakly erect or somewhat climbing evergreen shrub with very fragrant white, pink or yellowish flowers, borne singly or 2 or 3 together. Widely grown in many forms in greenhouses and gardens.

R. chinensis; China. Similar to above but with red, pink or whitish non-fragrant flowers. Some of the dwarf varieties have been developed from this.

Climbing Rose Species

Cherokee Rose; *R. laevigata*; China. Very vigorous, evergreen climber with stout stems and very stout hooked prickles; leaflets usually 3; flowers solitary, white and fragrant.

Lady Banksia Rose; *R. banksiae*; China. High-climbing evergreen with few or no prickles; leaflets 3-7; flowers white to yellow, about 1′ wide and in clusters, single or double.

Rose Acacia

Robinia hispida

Legume Family

GROWTH HABIT: Deciduous shrub to 7′ high, twigs bristly-hispid; colony forming
LEAVES: Alternate, pinnate with 7-13 oval entire leaflets
FLOWERS: Rose-colored, 1″ long and in short racemes
FRUITS: Pod 2-3″ long and hispid, seldom developed
SEEDS: Few
LIGHT: Full or partial sun
SOIL: Fertile, porous type
MOISTURE: Moist to dry
SEASONAL ASPECT: Flowers in late spring
ZONE: 1, 2, 3
USES: Bed, bank or wherever colony size is controllable
CARE: None
PROPAGATION BY: Stolons or root sprouts
NATIVE OF: s.e. United States
OTHER: Four or five other shrubby but less showy species are natives. *R. nana,* also a colony former is the least, seldom over 2′ high.

Rosemary

Rosmarinus officinalis

Mint Family

GROWTH HABIT: 2-3′ high evergreen shrub
LEAVES: Opposite, aromatic, thick, dotted, linear, about 1″ long and hairy beneath
FLOWERS: Light blue, 1/2″ long and in axillary spikes
FRUITS: 4 nutlets per flower
SEEDS: 1 per nutlet
LIGHT: Full or lots of sun
SOIL: Porous, fertile type
MOISTURE: Moist
SEASONAL ASPECT: Early spring flowers
ZONE: 1, 2
USES: Border, interest, herb garden
CARE: None
PROPAGATION BY: Seedlings
NATIVE OF: Mediterranean region
OTHER: Leaves used in seasoning and to furnish a volatile oil. Said to grow better around those homes where the mistress is master.

Rush

Juncus effusus

Rush Family

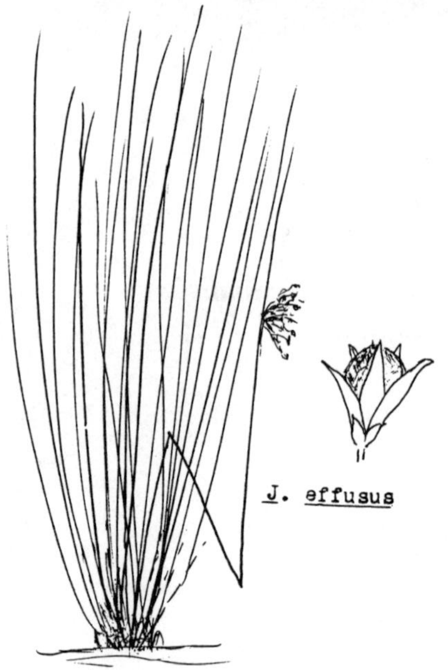

GROWTH HABIT: Clump-forming evergreen to 3′ high
LEAVES: Rudimentary on basal sheaths or lacking; the green stems carry on photosynthesis
FLOWERS: Small and borne near top of leafless stems; perianth parts persistent
FRUITS: Capsule
SEEDS: Very small and numerous
LIGHT: Full sun
SOIL: Heavy silt or clay loam
MOISTURE: Wet or moist
SEASONAL ASPECT: None
ZONE: 1, 2, 3
USES: Open wet borders
CARE: None
PROPAGATION BY: Clump division
NATIVE OF: e. United States
OTHER: *J. coriaceus* smaller and more diffuse clump former thriving in partial shade.

Russian Olive

Elaeagnus spp.

Oleaster Family

GROWTH HABIT: Evergreen and deciduous shrubs with brownish or silvery and often thorny twigs

LEAVES: Alternate, thickly covered, particularly beneath, with silvery or brownish scales

FLOWERS: Axillary, fragrant in some; calyx petaloid, tubular, 4-lobed and cream or yellow

FRUITS: Juicy pulp, pinkish acrid when ripe

SEEDS: Bony seed

LIGHT: Full or partial sun

SOIL: Tolerant of most types

MOISTURE: Moist

SEASONAL ASPECT: None

ZONE: 1, 2, 3

USES: Occasional shrub, hedge, screen

CARE: Pruning

PROPAGATION BY: Seedlings and cuttings

NATIVE OF: Japan, China

OTHER: *E. pungens* and varieties; China; evergreen, twigs densely covered with brownish scales and usually thorny; fall flowering.

E. macrophylla; Japan, Korea; evergreen; twigs densely covered with silvery scales.

E. umbellata; Orient; deciduous, twigs and undersurface of leaves covered with silvery-white scales; spring flowering; fruit edible as jelly and favored by birds.

Sageretia

Sageretia minutiflora

Buckthorn Family

GROWTH HABIT: Much-branched, thorny, deciduous shrub to 8′
LEAVES: Opposite, ovate, toothed and to 1 1/2″ long
FLOWERS: Minute, greenish and crowded in short axillary and terminal spikes
FRUITS: Dark purple drupes 1/4″ wide
SEEDS: 1 or 2
LIGHT: Full sun, or nearly so
SOIL: Sandy, enriched with lime
MOISTURE: Dry to moist
SEASONAL ASPECT: Fruits persisting into winter
ZONE: 1
USES: Specimen
CARE: None, or prune for shape
PROPAGATION BY: Seeds, cuttings
NATIVE OF: Coastal S.C. and south
OTHER: A very rare plant needing the protection of cultivation.

St. Andrew's Cross

Ascyrum hypericoides

St. Johnswort Family

GROWTH HABIT: Much-branched, semievergreen shrub to 30″ high
LEAVES: Opposite, linear to oblanceolate, glabrous and to 1″ long
FLOWERS: 4-parted, petals yellow and arranged as a St. Andrew's cross, two large and two small sepals
FRUITS: Capsule
SEEDS: Tiny
LIGHT: Full sun or partial shade
SOIL: Rather sterile
MOISTURE: Dry
SEASONAL ASPECT: Mid and late summer flowers
ZONE: 1, 2, 3
USES: Rock garden, barren spot, bank, thin woods
CARE: None
PROPAGATION BY: Seedlings
NATIVE OF: e. United States
OTHER: St. Peters-wort, *A. stans* is a larger leaved and larger flowered native and restricted to moist or wet places. Its flowers are also 4-parted.

Salt Bush

Atriplex breweri

Goosefoot Family

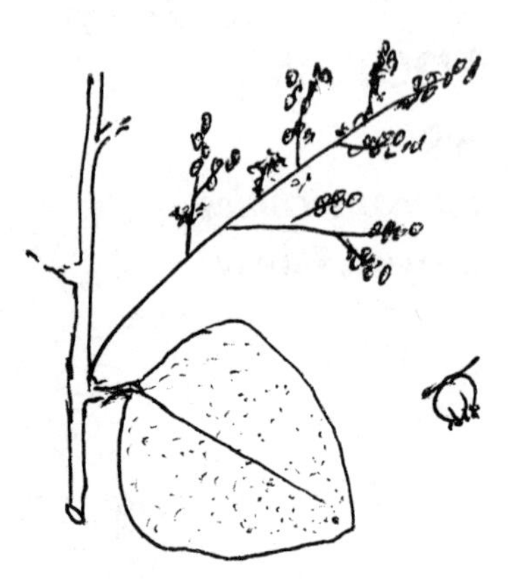

GROWTH HABIT: Erect much-branched semievergreen shrub to 8′ tall
LEAVES: Gray-scurfy, ornate and about 1″ long
FLOWERS: Male and female separate on same or different plants, small and in gray-green bracted spikes
FRUITS: Small and dry
SEEDS: 1
LIGHT: Full sun
SOIL: Sandy
MOISTURE: Moist
SEASONAL ASPECT: None
ZONE: 1, coastal
USES: Hedge
CARE: None
PROPAGATION BY: Seeds
NATIVE OF: California
OTHER: Good beach plant because of high salt tolerance.

Salvia

Salvia leucantha

Mint Family

GROWTH HABIT: Deciduous shrub to 2′ high with young branches white-wooly

LEAVES: 2-5″ long, narrowly lanceolate, rough pubescent and shallowly toothed

FLOWERS: White, 1/2″ or more long, subtended by calyx that is densely covered with violet hairs

FRUITS: 4 nutlets per flower

SEEDS: 1 seed per nutlet

LIGHT: Full sun

SOIL: Porous and fertile

MOISTURE: Dry to moist

SEASONAL ASPECT: Summer flowers

ZONE: 1, 2

USES: Specimen

CARE: None

PROPAGATION BY: Seedlings

NATIVE OF: Mexico

OTHER: *S. leucophylla* (*Andibertia nivia*); California; shrubby-based hoary subshrub to 3′ high; leaves to 2″ long; flowers bluish purple and 1/2″ long.

Sand Myrtle

Leiophyllum buxifolium

Heath Family

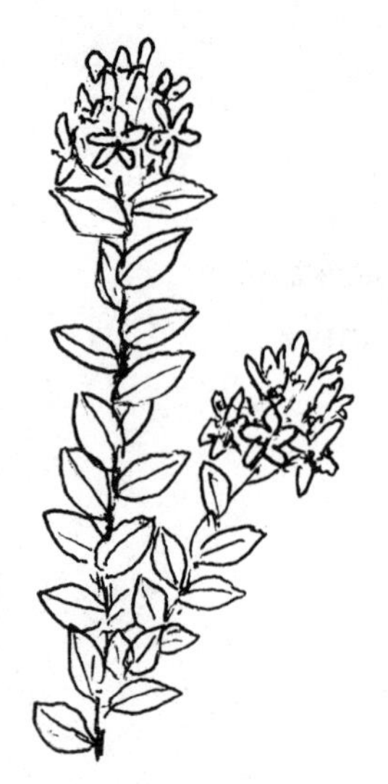

GROWTH HABIT: Spreading evergreen shrub to 2 1/2 ' high
LEAVES: Mostly alternate, about 1/4″ long, thick and shining above, Boxwood-like
FLOWERS: White to pinkish, 5-parted, barely 1/4″ wide
FRUITS: Capsule
SEEDS: Many
LIGHT: Almost full sun
SOIL: Very sandy
MOISTURE: Moist
SEASONAL ASPECT: Flowers in spring
ZONE: 1, 2, 3
USES: Bed, several together in sandy spot
CARE: Transplant very carefully
PROPAGATION BY: Transplants
NATIVE OF: e.s. United States

Sandwort

Arenaria caroliniana (Minuartia, Alsine)

Pink Family

GROWTH HABIT: Low evergreen with many short basal branches forming a cushion a few inches wide from a large deep tap root
LEAVES: Opposite, small, very narrow and sharp pointed
FLOWERS: White, 1/2″ wide and raised 6″ or more on forking stems
FRUITS: 3-valved capsule
SEEDS: Several
LIGHT: Full sun
SOIL: Sand
MOISTURE: Dry
SEASONAL ASPECT: Late spring flowers
ZONE: 1, 2
USES: Border, rock work, bed
CARE: None
PROPAGATION BY: Seeds
NATIVE OF: s.e. United States
OTHER: Several other species are cultivated in more northern latitudes.

Sarcococca

Sarcococca confusa

Box Family

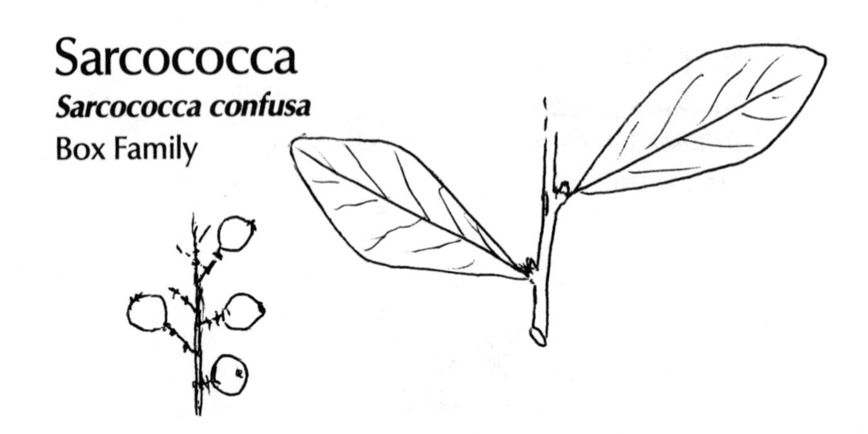

GROWTH HABIT: Evergreen shrub to 6′ high
LEAVES: Alternate, entire, elliptic, glossy and to 2″ long
FLOWERS: Inconspicuous and separate on same plant, fragrant
FRUITS: Leathery, berry-like
SEEDS: 1-2
LIGHT: Full or partial sun
SOIL: Fertile
MOISTURE: Moist
SEASONAL ASPECT: None
ZONE: 1, 2, 3
USES: Foundation, shrub border, interest, patio
CARE: Some clipping for desired shape
PROPAGATION BY: Cuttings
NATIVE OF: China
OTHER: A vigorous grower.
S. *ruscifolia*; China; broad leaves and red berries.
S. *hookeriana*; China; narrow leaves and black berries.

Privet, Evergreen species

Ligustrum spp.

Olive Family

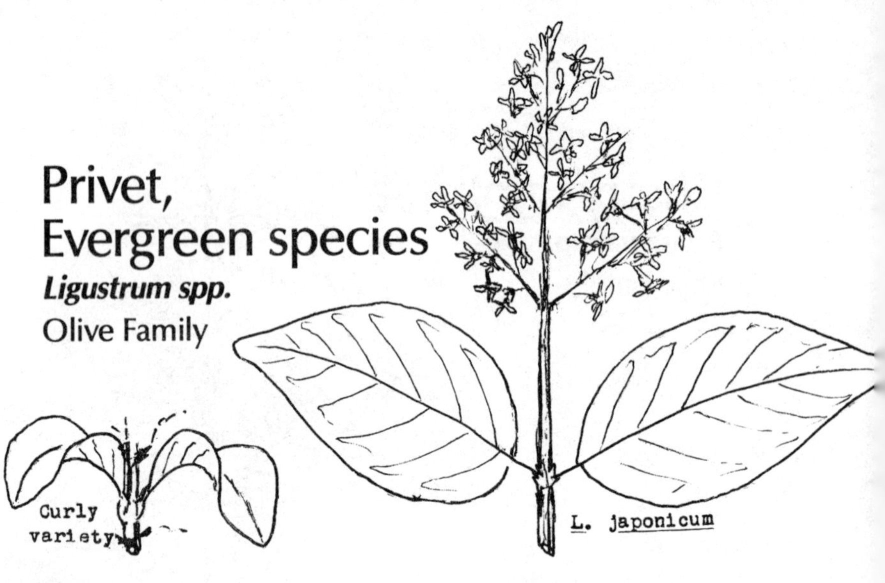

GROWTH HABIT: Evergreen shrubs extensively used and extremely hardy. All species below grown in several varieties

LEAVES: Opposite, entire, glabrous and 1-5″ long

FLOWERS: Small, white, numerous and in 2-8″ long panicles

FRUITS: Small, black berries

SEEDS: 1

LIGHT: Sun or shade

SOIL: Any type

MOISTURE: Wet, moist or dry

SEASONAL ASPECT: Late spring or early summer flowers

ZONE: 1, 2, 3

USES: Hedge, border, background, screen

CARE: Pruning

PROPAGATION BY: Cuttings

NATIVE OF: China, Japan, Korea

OTHER: Glossy Privet; *L. lucidum*; evergreen; leaves 3 1/2 - 5″ long.

Japanese Privet; *L. japonicum*; evergreen; leaves 2-3″ long and very dark green; similar to above; both often escaped. A densely foliated, curly leaf variety is grown.

California Privet; *L. ovalifolium*; evergreen with us; leaves 1 - 2 1/2″ long, dark green above and yellowish beneath; leaves and twigs glabrous; panicles erect and to 4″ long.

Border Privet; *L. obtusifolium*; evergreen with us; leaves 1-2″ long, pubescent on midrib beneath; panicles 1 1/2″ long and nodding.

Sassafras

Sassafras albidum

Laurel Family

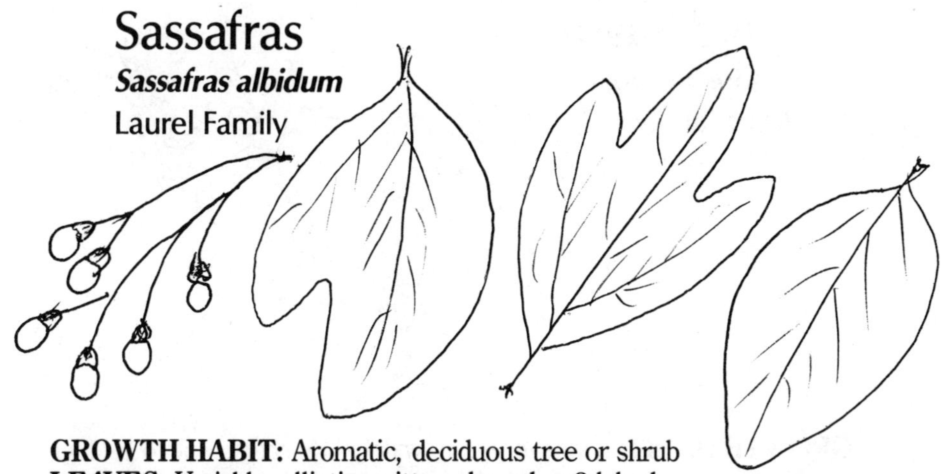

GROWTH HABIT: Aromatic, deciduous tree or shrub
LEAVES: Variable; elliptic, mitten-shaped or 3-lobed
FLOWERS: Small, greenish-yellow, from branch tips before the leaves; male and female on separate plants
FRUITS: Blue, 1/2″ long on elongate pedicel enlarged and red at tip
SEEDS: 1
LIGHT: Lots of or full sun
SOIL: Sandy
MOISTURE: Dry
SEASONAL ASPECT: Autumn colors
ZONE: 1, 2, 3
USES: Freestanding, interest
CARE: None
PROPAGATION BY: Seedlings
NATIVE OF: e. United States
OTHER: Roots formerly popular as a source of tea and with birch bark, (*Betula lenta*) the source of flavoring for root beer.

Wild Savory

Satureja georgiana

Mint Family

GROWTH HABIT: Semievergreen, widely-branched shrub to 18″ high with pubescent twigs
LEAVES: Opposite, elliptic or wider, to about 1″ long and entire, toothed or with irregular margin
FLOWERS: Lavender, pink or white
FRUITS: 4 nutlets per flower
SEEDS: 1 per nutlet
LIGHT: Partial sun
SOIL: Tolerant to most types
MOISTURE: Moist or dry
SEASONAL ASPECT: Late summer flowers
ZONE: 1, 2
USES: Specimen, bank, margin
CARE: None
PROPAGATION BY: Seedlings
NATIVE OF: s.e. United States

Scholar-Tree, Japanese Pagoda Tree

Sophora japonica

Legume Family

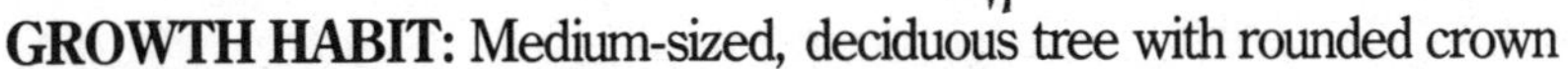

GROWTH HABIT: Medium-sized, deciduous tree with rounded crown
LEAVES: Alternate, pinnate, to 9″ long and 7-17 leaflets
FLOWERS: 1/2″ long, creamy-white; pea-shaped; in large panicles
FRUITS: 2-3″ long, smooth pod
SEEDS: Few
LIGHT: Hot sunny place
SOIL: Moist types suitable
MOISTURE: Dry
SEASONAL ASPECT: Late summer flowers
ZONE: 1, 2
USES: Street, sidewalk, lawn
CARE: None
PROPAGATION BY: Seeds and seedlings
NATIVE OF: China
OTHER: Weeping form available.

Scotch Broom

Cytissus scoparius

Legume Family

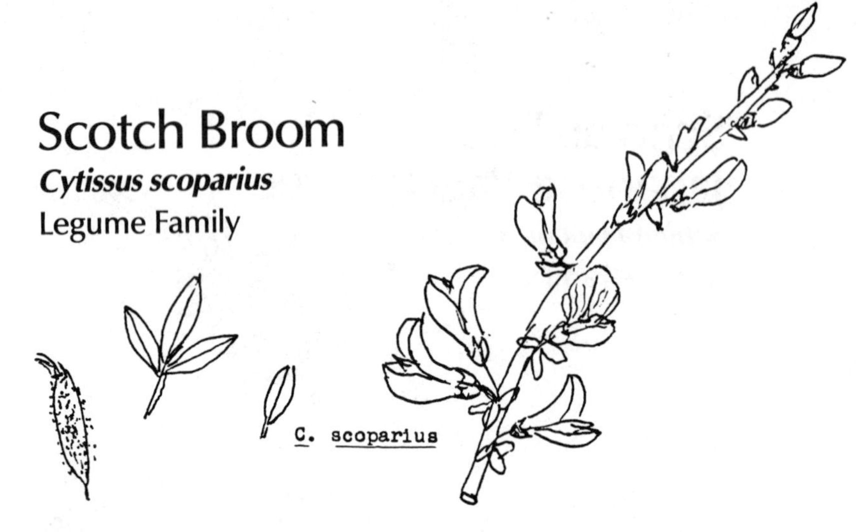

GROWTH HABIT: Bushy, deciduous shrub to 5′ high, twigs and branches green

LEAVES: Alternate, short petioled, mostly 3 foliate, leaflets entire and less than 1″ long

FLOWERS: Yellow, 3/4″ long and produced abundantly along the stems

FRUITS: Dark colored 1 1/2″ long pods

SEEDS: 2-5

LIGHT: Full sun

SOIL: Tolerant

MOISTURE: Dry to moist

SEASONAL ASPECT: Late spring flowers

ZONE: 1, 2, 3

USES: Bank, margin, occasional flowering shrub, accent

CARE: None

PROPAGATION BY: Seedlings

NATIVE OF: Central and south Europe, sometimes escaped in our area

OTHER: A number of other yellow-flowered species and hybrids, mostly of *C. scoparius*, are offered as well as the following:

C. albus; a 1′ high, white flowered form from Portugal.

C. purpureus; an 18″ high, purple-flowered form from s. Europe.

Sea Oxeye

Borrichia frutescens

Composite Family

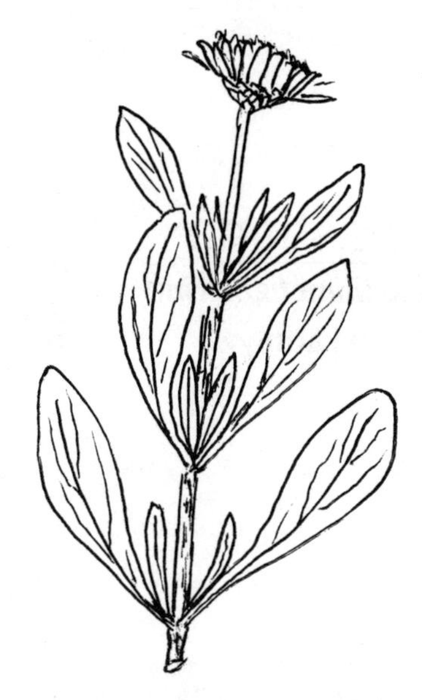

GROWTH HABIT: Tardily deciduous shrub to 2′ high, little branched and strongly stoliniferous
LEAVES: Opposite, oblanceolate, mostly entire and hairy
FLOWERS: Rays yellow, disk brownish
FRUITS: Small achene
SEEDS: 1
LIGHT: Full sun
SOIL: Marsh and ocean margins
MOISTURE: Wettish
SEASONAL ASPECT: Summer flowers
ZONE: 1
USES: In brackish marsh situations or high tide lines
CARE: None
PROPAGATION BY: Stolons
NATIVE OF: s.e. United States coast

Sebastiana

Sebastiana ligustrina
Spurge Family

GROWTH HABIT: Deciduous shrub to 6′ high
LEAVES: Alternate, elliptic, entire, to 2 1/2″ long
FLOWERS: Small, green, in racemes, male above, female few and below
FRUITS: 1/2″ long 3-celled capsule
SEEDS: 3
LIGHT: Shifting shade from canopy
SOIL: Sandy
MOISTURE: Dry to moist
SEASONAL ASPECT: None
ZONE: 1
USES: Specimen
CARE: None
PROPAGATION BY: Seeds
NATIVE OF: s.e. United States coastal plains

Sedge

Carex spp.

Sedge Family

Perhaps as many as 150 species of this large and common genus are native to this area. They are perennials, mostly tufted and with narrow grass-like leaves borne in three ranks. Some are evergreen and most of these are less than 1′ high. Flowers are small, green and somewhat grass-like. Male and female are separate and either on separate spikes or separate on the same spike. The fruits are either 2- or 3-sided achenes each enclosed in a small sac-like structure that may be close or loose-fitting. Identification to species is difficult, but desirable specimens may be collected and used in borders, as edging or as interesting and different bits of greenery.

Commonly mistaken for some kind of grass.

Serissa

Serissa foetida

Madder Family

GROWTH HABIT: Small, semievergreen shrub with slender branches
LEAVES: Opposite, 1/2″ long and leathery
FLOWERS: White, from pink buds and to almost 1/2″ long
FRUITS: Small capsule
SEEDS: 2
LIGHT: Partial shade
SOIL: Fertile loam
MOISTURE: Moist
SEASONAL ASPECT: None
ZONE: 1, 2
USES: Terrace, patio, pot
CARE: Protect from cold
PROPAGATION BY: Cuttings
NATIVE OF: Japan
OTHER: A variety with yellow-margined leaves exists.

Service-Berry, Shadblow, June-Berry

Amelanchier arborea

Rose Family

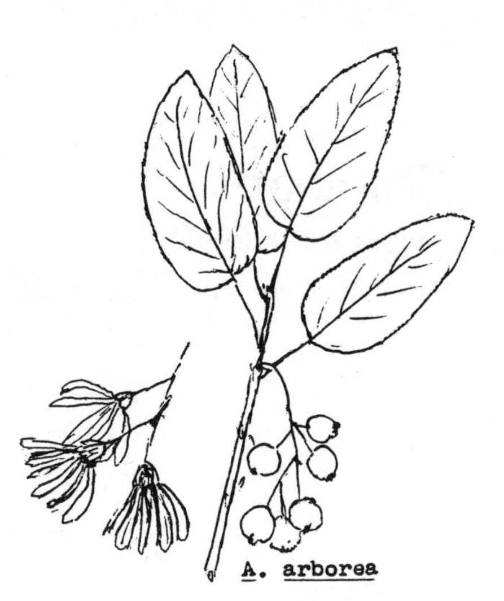

GROWTH HABIT: Deciduous shrubs or occasionally small trees
LEAVES: Alternate, more-or-less ovate and finely serrate
FLOWERS: White, medium-sized and in short drooping racemes
FRUITS: Purple, juicy, berry-like and delicious
SEEDS: Several and small
LIGHT: Some shade
SOIL: Fertile
MOISTURE: Moist but well drained
SEASONAL ASPECT: Showy white flowers in very early spring
ZONE: 1, 2, 3
USES: As large shrub or tree
CARE: Maybe some pruning
PROPAGATION BY: Seedlings
NATIVE OF: e. United States
OTHER: *A. canadensis*; e. United States; very similar but has erect racemes.

Shinleaf

Pyrola rotundifolia

Heath Family

GROWTH HABIT: Evergreen subshrub to 1′ high
LEAVES: Alternate, basal, leathery and to 2″ wide
FLOWERS: White, 5-parted and raised above the leaves
FRUITS: Capsule
SEEDS: Many and tiny
LIGHT: Shade
SOIL: Humus rich
MOISTURE: Moist
SEASONAL ASPECT: Summer flowers
ZONE: (2), 3
USES: Specimen
CARE: None
PROPAGATION BY: Transplants
NATIVE OF: n.e. United States, s. to Ga.

Shrub Plumbago

Ceratostigma willmottianum

Leadwort Family

GROWTH HABIT: Deciduous shrub to 4′ high with angled, bristly, somewhat purplish stems
LEAVES: Alternate, lanceolate to obovate to 2″ long and hairy beneath
FLOWERS: 1/2″ long, petals blue, tube red, in terminal heads with bracts between the flowers
FRUITS: Small capsule
SEEDS: Several
LIGHT: Lots of sun
SOIL: Moderately fertile
MOISTURE: Moist but well drained
SEASONAL ASPECT: Late summer and fall flowers
ZONE: 1, 2
USES: Specimen, interest
CARE: None
PROPAGATION BY: Seedlings, cuttings
NATIVE OF: China

Carolina Silver-Bell

Halesia carolina

Storax Family

GROWTH HABIT: Deciduous shrub or small tree with conspicuous striped bark until 3-4″ in diameter
LEAVES: Alternate, ovate, accuminate, 2″- long and finely toothed; pubescent beneath
FLOWERS: White, bell-shaped, 1/2-1″ long, drooping
FRUITS: 1 - 1 1/2″ long, nut-like and with 4 longitudinal wings
SEEDS: 1
LIGHT: Partial shade
SOIL: Fertile loam rich in humus
MOISTURE: Moist
SEASONAL ASPECT: Spring flowers just as leaves appear
ZONE: (1), 2, 3
USES: Background, specimen
CARE: None
PROPAGATION BY: Seeds and seedlings
NATIVE OF: e. United States
OTHER: *H. diptera*; lower coastal plain, S.C.-Miss.; similar to above but with 2-winged fruit.
H. parviflora; S.C.-Ala.; similar but with flowers less than 1/2″ long.

Skimmia

Skimmia japonica

Rue Family

GROWTH HABIT: Dense, evergreen shrub, aromatic and slow growing
LEAVES: Alternate, obovate to elliptic, leathery
FLOWERS: White, 1/2″ wide, clustered, male and female on separate plants
FRUITS: Bright red, 1/4″ wide
SEEDS: Few
LIGHT: Lots of sun
SOIL: Any fertile, porous type
MOISTURE: Moist to dry
SEASONAL ASPECT: None
ZONE: 1, 2, 3
USES: Border, unit arrangement, specimen, industrial area, seaside garden
CARE: Very little pruning
PROPAGATION BY: Cuttings in summer
NATIVE OF: China
OTHER: A number of varieties have been developed. Fruit toxic.

Smoke-Tree

Cotinus obovatus (americanus)

Cashew Family

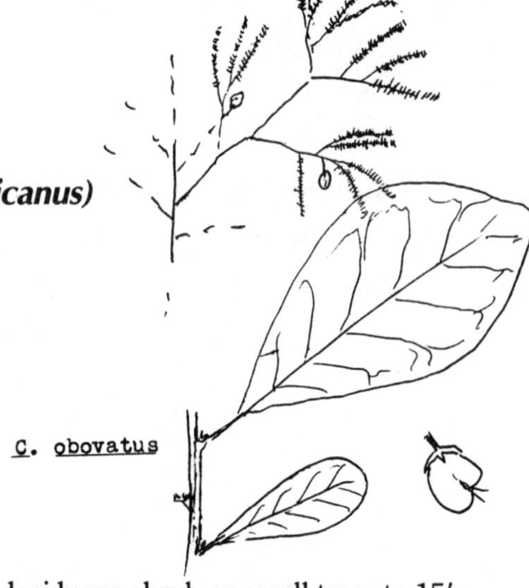

GROWTH HABIT: Spreading, deciduous shrub or small tree, to 15′
LEAVES: Alternate, obovate, 1-3″ long and entire
FLOWERS: Mostly abortive, sterile pedicels becoming much elongated, purplish and plumose; fertile ones short, glabrous and supporting a very small greenish flower
FRUITS: Somewhat kidney shaped and less than 1/2″ wide
SEEDS: 1 or 2
LIGHT: Some shade, perhaps half
SOIL: Good loose fertile type
MOISTURE: Moist
SEASONAL ASPECT: Flowering time in summer and orange to scarlet autumn colors
ZONE: 3
USES: Specimen
CARE: None
PROPAGATION BY: Cuttings
NATIVE OF: e. United States
OTHER: This is cultivated in several varieties. *C. coggygria*; Europe, China; fruiting branches more showy; leaves smaller.

Snow-Wreath

Neviusia alabamensis

Rose Family

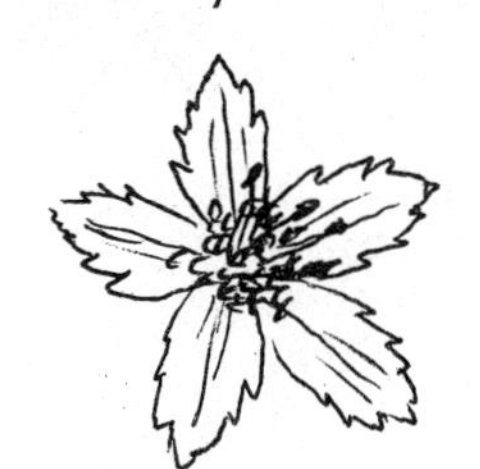

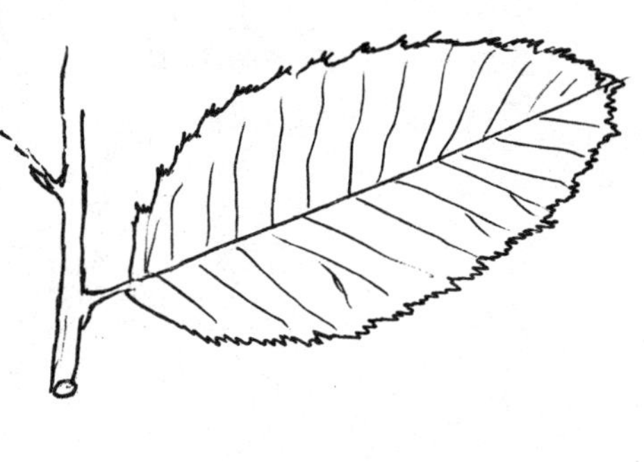

GROWTH HABIT: Hardy, deciduous shrub to 5′ with slender branching stems
LEAVES: Alternate, doubly toothed
FLOWERS: No petals; sepals petal-like and whitish; stamens white
FRUITS: Rather large achene
SEEDS: 1
LIGHT: Lots of sun
SOIL: Porous
MOISTURE: Dry
SEASONAL ASPECT: Spring flowers
ZONE: 1, 2, 3
USES: Bed, border, woodland, margin
CARE: Some pruning to keep stems from overcrowding
PROPAGATION BY: Cuttings, seedlings
NATIVE OF: Alabama

Soapberry

Sapindus drummondii

Soapberry Family

GROWTH HABIT: Deciduous tree
LEAVES: Alternate, pinnate, leaflets entire and lanceolate to somewhat sickle-shaped
FLOWERS: Small, greenish and in large panicles
FRUITS: Globose, 1/2″ wide and usually showing two very small abortive fruits adhering near the point of attachment
SEEDS: 1
LIGHT: Full sun
SOIL: Tolerant of most productive types
MOISTURE: Moist to dry
SEASONAL ASPECT: None
ZONE: 1, 2
USES: Interest
CARE: None
PROPAGATION BY: Seedlings
NATIVE OF: central southern United States

Sourwood

Oxydendrum arboreum

Heath Family

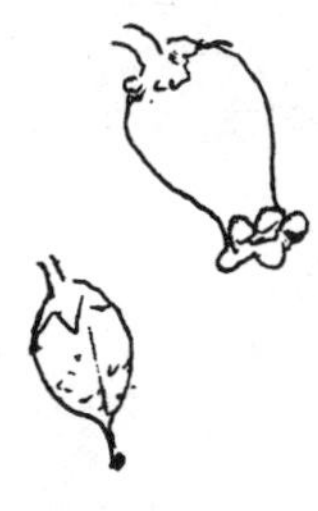

GROWTH HABIT: Deciduous tree with red twigs, light gray checkered bark and a leaning or stooping habit
LEAVES: Alternate, elliptic to lanceolate, finely and sharply toothed, sour tasting and to 7″ long
FLOWERS: White, 1/4″ long, urn-shaped and in clustered terminal racemes
FRUITS: A small woody capsule
SEEDS: Several and tiny
LIGHT: Partial sahde
SOIL: Tolerant of most acid types
MOISTURE: Moist
SEASONAL ASPECT: Autumn foliage
ZONE: (1), 2, 3
USES: Background, freestanding, interest, fall color
CARE: None
PROPAGATION BY: Seedlings
NATIVE OF: e. United States
OTHER: Sometimes in the trade as Lily-of-the-Valley-Tree. The source of sourwood honey. Shows autumn color early.

Spanish Bayonet, Yucca

Yucca aloifolia

Lily Family

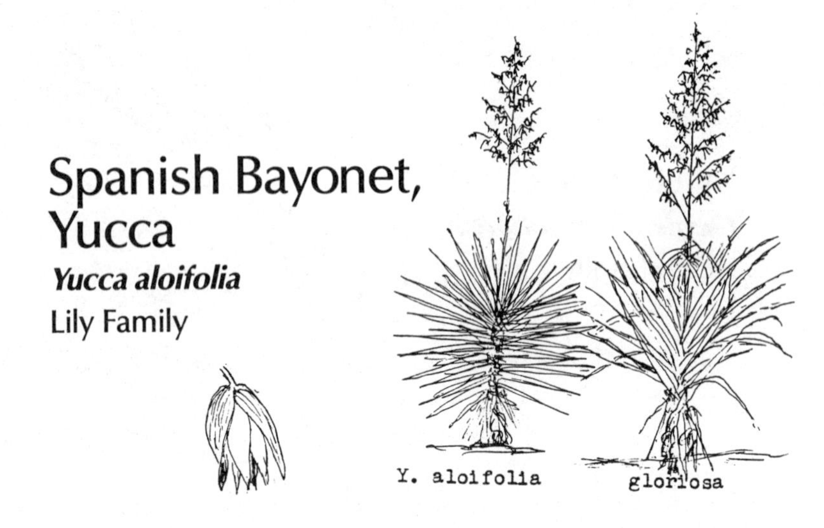

GROWTH HABIT: Rosette and eventually trunk-forming evergreen, trunk sometimes branched and to several feet high

LEAVES: Stiff and dagger-shaped with very sharp points and sharply toothed margins, to 2″ wide and 3′ long

FLOWERS: White, 2″ long, hanging, waxy on large panicles well overtipping the leaves

FRUITS: Somewhat fleshy capsule to 4″ long

SEEDS: Many and turgid

LIGHT: Full sun

SOIL: Most any type

MOISTURE: Moist or dry

SEASONAL ASPECT: Fall flowers

ZONE: 1, 2

USES: Barrier, specimen, background

CARE: None

PROPAGATION BY: Seedlings, trunk sections

filamentosa

NATIVE OF: s.e. United States coastal area

OTHER: *Y. gloriosa*, s.e. United States coastal area. Similar but leaf margins entire but sharp-edged; plant larger and darker green.

Bear-Grass; *Y. filamentosa*; s.e. United States with no trunk and flexible leaves, the margins fraying into filaments; wide and narrow leaved varieties, some variegated; root sprouts are common. The leaves, strong and fibrous, were formerly of some use to country folk for hanging meat in smoke houses.

Spanish Broom

Genista hispanica

Legume Family

GROWTH HABIT: Low, deciduous shrub, but green branches and large green spines give it an evergreen appearance; to 1 1/2′ high
LEAVES: Broadly lanceolate and 1/2″ long
FLOWERS: Bright yellow 1/2″ long and in clusters
FRUITS: Short pod
SEEDS: 1-2
LIGHT: Will stand some shade
SOIL: Favor sandy
MOISTURE: Dry
SEASONAL ASPECT: Late spring flowers
ZONE: 1, 2, 3
USES: Rockery or other dry place, barrier
CARE: None
PROPAGATION BY: Seeds
NATIVE OF: Spain to s. Italy

Spanish Moss

Tillandsia usneoides

Pineapple Family

GROWTH HABIT: Evergreen epiphyte, festooning rootless plant suspended from tree branches and deriving its water and minerals from the air
LEAVES: 1-2″ long from almost filamentous stems and densely covered with gray scurfy scales
FLOWERS: Small; petals 3 and greenish
FRUITS: Capsule 1″ long and splitting into 3 parts
SEEDS: Minute and with silky appendage
LIGHT: Partial or full sun
SOIL: None
MOISTURE: Coastal and bottom land forests
SEASONAL ASPECT: None
ZONE: 1, 2
USES: Interest
CARE: None
PROPAGATION BY: Seeds
NATIVE OF: Coastal s.e. United States and inland along waterways

Sparkleberry

Vaccinium arboreum

Heath Family

GROWTH HABIT: Much-branched evergreen or tardily-deciduous shrub to 10′ high
LEAVES: Leathery, elliptic, oval to obovate, entire and to 1 or 1 1/2″ long
FLOWERS: Numerous, white, 1/4″ long with short but distinct lobes
FRUITS: Black, lustrous, mealy, rather sweet 1/4″ long or less
SEEDS: Many
LIGHT: Partial sun
SOIL: Sandy, mostly soils of low fertility
MOISTURE: Dry
SEASONAL ASPECT: Spring flowers
ZONE: 1, 2, 3
USES: Occasional shrub, background, interest
CARE: None
PROPAGATION BY: Seedlings
NATIVE OF: e. United States

Spice-Bush

Lindera benzoin

Laurel Family

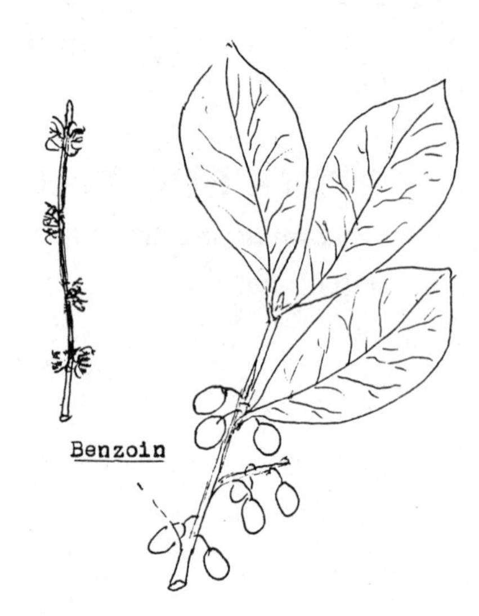

GROWTH HABIT: Much-branched, deciduous shrub to 12′ high, aromatic

LEAVES: Alternate, elliptic to obovate, entire, leaves at twig tips usually largest

FLOWERS: Male and female flowers usually on separate plants, before the leaves, small, yellow and in axillary clusters

FRUITS: 1/3″ long, bright red and very aromatic when crushed

SEEDS: 1

LIGHT: Partial or full shade

SOIL: Silt or silty loam

MOISTURE: Moist

SEASONAL ASPECT: None

ZONE: 1, 2, 3

USES: Interest, occasional shrub

CARE: None

PROPAGATION BY: Seedlings and cuttings

NATIVE OF: s.e. United States

OTHER: *Litsea aestivalis* is a related deciduous shrub; lower coastal plain from North Carolina to Florida. It is native to low, wet woodlands, has red drupes and is one of our rarest shrubs; occasionally cultivated for interest and conservation.

Spiderwort

Tradescantia ohiensis

Spiderwort Family

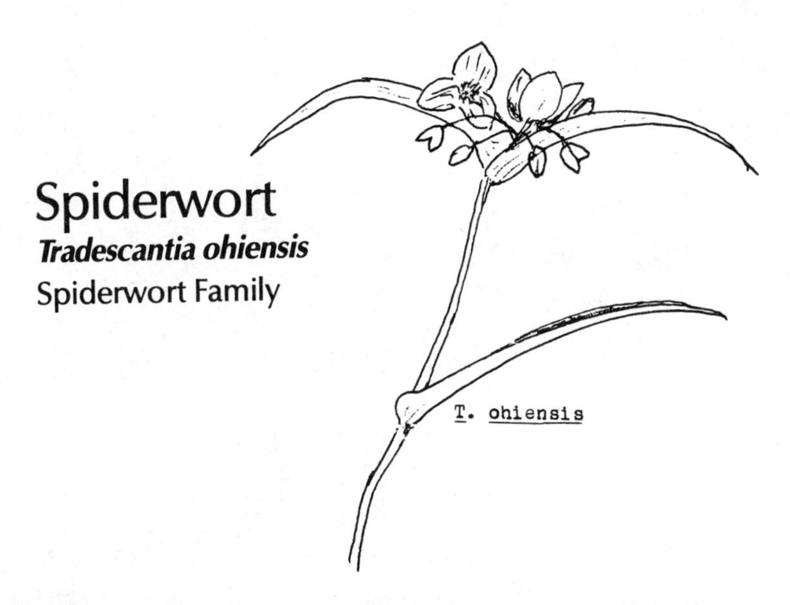

GROWTH HABIT: Nearly evergreen herbaceous tufted plant to 2′ high from stout fleshy roots
LEAVES: Linear, to 1 1/2′ long and 1/2″ wide
FLOWERS: Blue to rose, 1″ wide, 3-parted with several flowers and buds topping leafy stem; showy and long blooming
FRUITS: Capsule
SEEDS: Several and small
LIGHT: Lots of sun
SOIL: Sandy loam
MOISTURE: Moist or dry
SEASONAL ASPECT: Spring and early summer flowers
ZONE: 1, 2, 3
USES: Interest
CARE: None
PROPAGATION BY: Seedlings or transplant
NATIVE OF: s.e. United States
OTHER: *T. hirsuticaulis*; s.e. United States; pubescent, more nearly evergreen plant; use in zones 2 and 3.
T. rosea; s.e. United States; Smaller plant with shorter narrower leaves and rose-colored flowers; used in sandy areas, zones 1, 2.

Spirea, Bridal-Wreath

Spiraea spp.

Rose Family

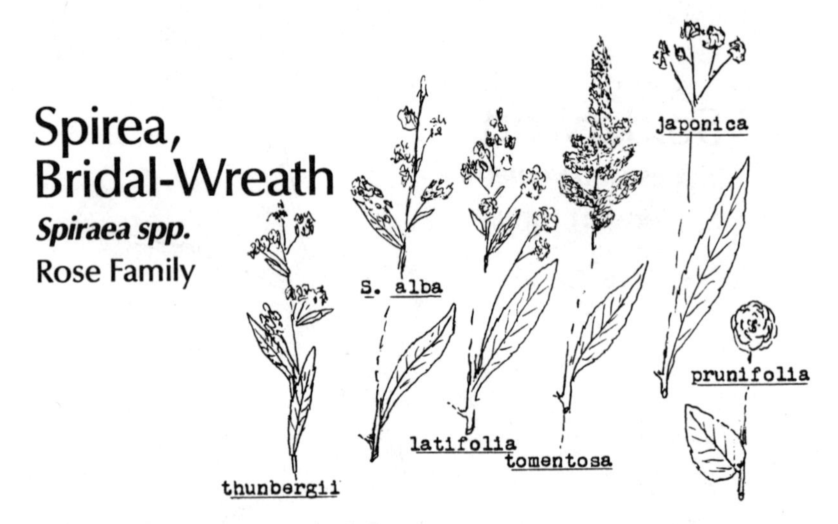

GROWTH HABIT: Rather small, deciduous shrubs with very slender twigs; many species and many varieties and hybrids
LEAVES: Alternate, small, usually toothed
FLOWERS: White, pink or purplish, borne singly or on clusters, single or double
FRUITS: 1-5 very small pods (follicles) per flower
SEEDS: Several and tiny
LIGHT: Lots of sun
SOIL: Tolerant of many types
MOISTURE: Dry to moist
SEASONAL ASPECT: Flowering time, spring and summer
ZONE: 1, 2, 3
USES: Border, unit arrangement, occasional flowering shrub
CARE: Some pruning
PROPAGATION BY: Cuttings
NATIVE OF: See below.
OTHER: Native species—mountains:
Meadow Sweet, S. *alba*; white flowers, white stamens. S. *latifolia*; white flowers; purple stamens. Steeplebush, S. *tomentosa*; pink flowers.

Introduced species:
S. *hypericoides*; s.e. Asia; dense bushy shrub with white flowers.
S. *japonica*; Japan; flowers pink and in clusters.
Bridal Wreath, S. *prunifolia*; Japan; flowers white and doubled, appearing after the leaves.
Early Spirea, S. *thumbergii*; China; very early white flowers; very narrow leaves.
S. *X vanhouttei* (hybrid between S. *cantoniensis* and S. *trilobata*) one of the most popular spring blooming types, rounded, densely packed flower clusters and toothed and lobed leaves.

Spotted Wintergreen, Pipsissewa

Chimaphila maculata

Heath Family

GROWTH HABIT: Evergreen subshrub to 8″ high from rhizomes
LEAVES: Alternate, lanceolate to narrowly elliptic, to 2″ long, toothed and mottled
FLOWERS: White, 3/4″ wide and raised above the leaves
FRUITS: Capsule
SEEDS: Very small
LIGHT: Partial shade
SOIL: Acid loam
MOISTURE: Dry to moist
SEASONAL ASPECT: Late spring flowers
ZONE: 1, 2, 3
USES: Interest
CARE: None
PROPAGATION BY: Plants with attached rhizomes
NATIVE OF: s.e. United States

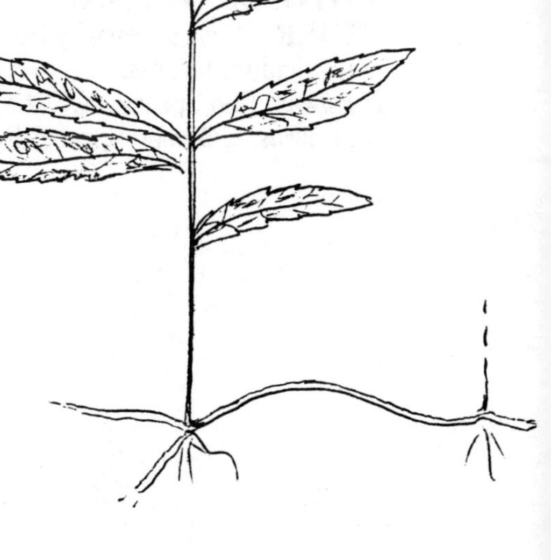

Spruce

Picea spp.

Pine Family

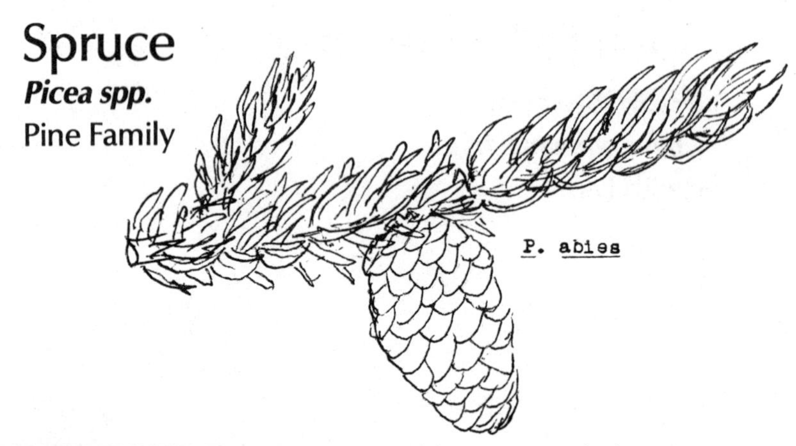

GROWTH HABIT: Evergreen trees with narrow conical crowns
LEAVES: Linear, spirally arranged, usually 4-angled and jointed at base to a short persistent leaf stalk
FLOWERS: Male as yellow or reddish catkins; female as cones
FRUITS: Pendulous woody cones
SEEDS: With large thin wing
LIGHT: Partial or full sun
SOIL: Clay loam
MOISTURE: Moist
SEASONAL ASPECT: None
ZONE: 3, intolerant to hot summers
USES: Freestanding, framing or background
CARE: None
PROPAGATION BY: Seedlings
NATIVE OF: See below
OTHER: *P. abies*, Norway Spruce, central and north Europe; in many horticultural forms.
P. rubens, Red Spruce; N.C. to N.S.
P. pungens, Colorado Spruce; Rocky Mountains from Wyoming south

Stachyurus

Stachyurus praecox

Stachyurus Family

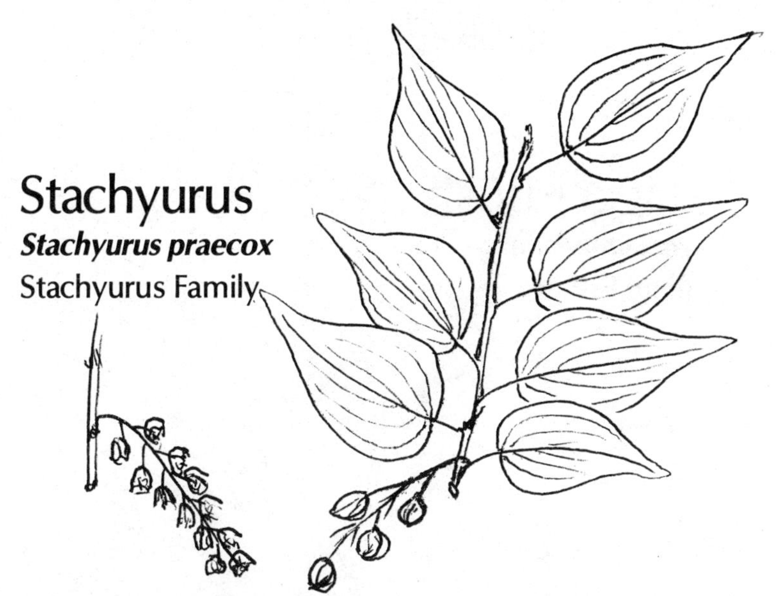

GROWTH HABIT: Tardily deciduous shrub to 8′ high
LEAVES: Alternate, ovate and long pointed, to 3″ long
FLOWERS: Small, greenish-yellow, 4-parted in 12-20 flowered racemes from leaf axils
FRUITS: 1/4″ wide berry
SEEDS: Usually 4
LIGHT: Partial sun
SOIL: Moderately fertile loam
MOISTURE: Dry
SEASONAL ASPECT: None
ZONE: 1, 2, 3
USES: Occasional shrub, interest
CARE: Some pruning after blooming
PROPAGATION BY: Seeds
NATIVE OF: Himalayas

Stagger-Bush

Lyonia mariana

Heath Family

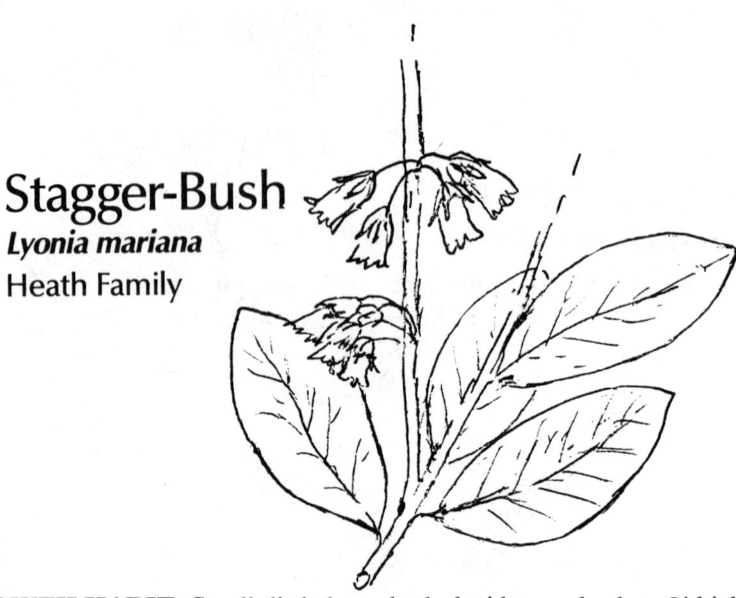

GROWTH HABIT: Small, little branched, deciduous shrub to 3′ high
LEAVES: Alternate, elliptic, entire, to 2 1/2″ long
FLOWERS: 3/4″ long, white or tinted, broadly urn-shaped and in small clusters
FRUITS: 1/4″ long capsule
SEEDS: Small and several
LIGHT: Three-fourths sun
SOIL: Sandy
MOISTURE: Moist
SEASONAL ASPECT: Spring flowers
ZONE: 1, 2
USES: Blender, low background, interest
CARE: None
PROPAGATION BY: Root sprouts or transplants
NATIVE OF: e. United States
OTHER: Slowly colony forming.

Stephanandra

Stephanandra incisa

Rose Family

GROWTH HABIT: Deciduous shrub with long arching branches to 6′ high
LEAVES: Alternate, triangular to ovate, long pointed, deeply lobed and toothed
FLOWERS: Very small, white, very numerous in axillary clusters
FRUITS: Tiny capsules
SEEDS: Tiny, if formed
LIGHT: Full or partial sun
SOIL: Fertile loam
MOISTURE: Moist
SEASONAL ASPECT: Spring flowers with or after leaves
ZONE: 2, 3
USES: Border, hedge, occasional flowering shrub
CARE: Prune early; flowers borne on new growth
PROPAGATION BY: Early summer cuttings and layering
NATIVE OF: Japan, China

Stewartia

Stewartia malacodendron

Tea Family

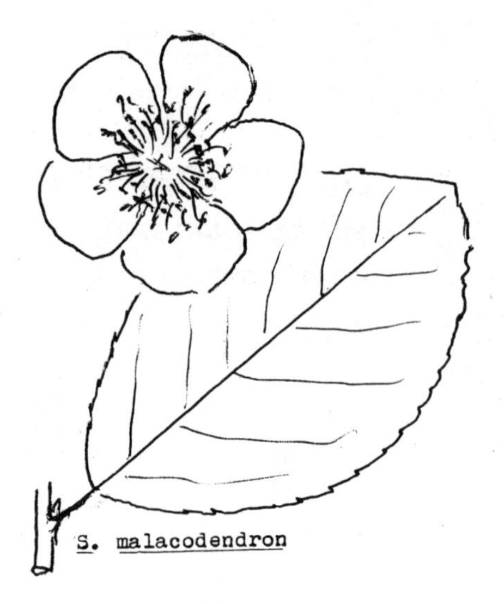

GROWTH HABIT: Large deciduous shrub

LEAVES: Alternate, elliptic, toothed, short pointed and pubescent beneath

FLOWERS: White, 2-3″ across, petals silky on outside; stamens numerous and purple; showy

FRUITS: Woody capsule 3/4″ long

SEEDS: Several and narrowly winged

LIGHT: Shifting shade from high canopy

SOIL: Fertile silt or clay loam

MOISTURE: Moist to wettish

SEASONAL ASPECT: Flowering time which is just after the leaves are full grown

ZONE: 1, 2

USES: Freestanding, interest shrub or to give a flash of color to a shrub border

CARE: None

PROPAGATION BY: Seedlings and cuttings

NATIVE OF: s.e. United States coastal plain

OTHER: S. *ovata,* s. Appalachians; similar to above but with white stamens; variety *grandiflora* produces flowers to 4″ wide.

S. *pseudocamellia;* tree of medium size.

S. *koreana* and S. *sinensis* are both shrub size and produce flowers to 2″ wide.

Stonecrop
Sedum spp.
Orpine Family

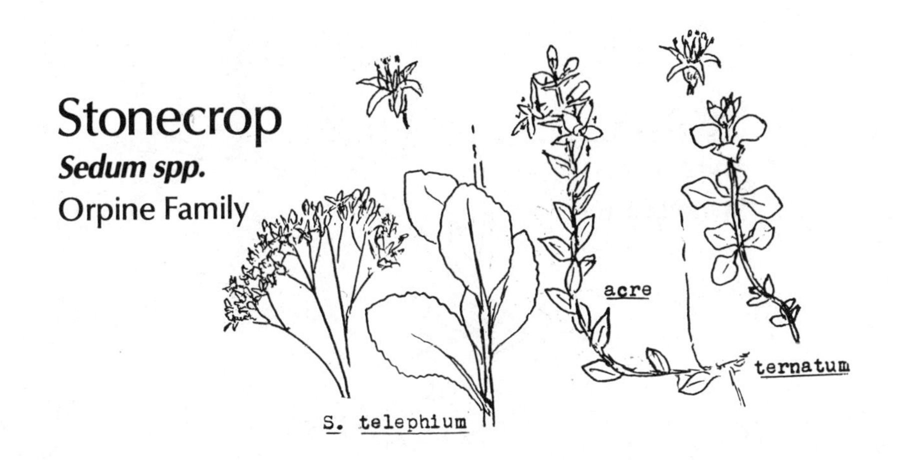

GROWTH HABIT: Succulent evergreen, or nearly evergreen, perennials or subshrubs, mostly low and mat-forming
LEAVES: Usually small, glabrous, alternate; and terete, or thick
FLOWERS: Small, regular, 4-5 parted, white, yellow or purple and in raised arrangements
FRUITS: 4-5 small pods or follicles per flower
SEEDS: Few per follicle
LIGHT: Full or partial sun
SOIL: Tolerant
MOISTURE: Moist to dry
SEASONAL ASPECT: None
ZONE: 1, 2, 3
USES: Around rock, stump, tree base, on bank
CARE: Occasional weeding until mat formation
PROPAGATION BY: Clumps or plant portions
NATIVE OF: see below
OTHER: Stonecrops—low mat-formers:
S. *acre*; Old World; leaves alternate and broad; flowers yellow.
S. *rupestre*; Portugal; leaves alternate; narrow and crowded toward stem tips; flowers yellow.
S. *spurium*; Caucasus; leaves opposite; flower white.
S. *ternatum*; e. North America; leaves whorled; flowers white.

Live-for-evers—succulent perennials to 2′ high:
S. *spectabile*; Japan; leaves opposite or whorled; flowers pink.
S. *telephioides*; e. United States; leaves alternate, upper and lower about same size; flowers white.
S. *telephium (purpureum)*; Europe to Japan; leaves alternate; upper much smaller; flowers deep pink or white.

Storax

Styrax spp.

Storax Family

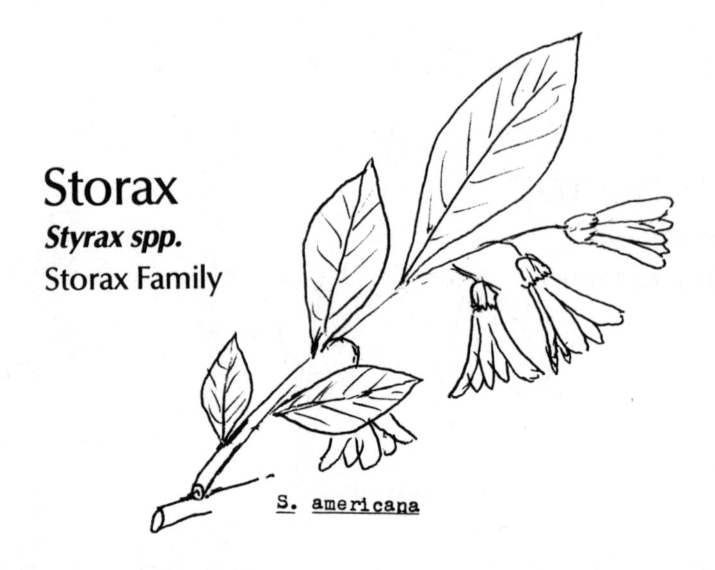

GROWTH HABIT: Deciduous shrubs to 10′ high
LEAVES: Alternate, elliptic to oval or ovate, usually acuminate, weakly toothed or entire and pubescent with stellate hairs
FLOWERS: White, fragrant 1/2 - 3/4″ long, drooping and in few to several flowered racemes
FRUITS: Dry, globose, 1/2″ long, 3-valved capsule
SEEDS: Usually 1
LIGHT: Partial to almost full sun
SOIL: Productive, usually sandy loam
MOISTURE: Moist to wet
SEASONAL ASPECT: Spring flowers with the leaves
ZONE: 1, 2, 3 (except over 3000 ft)
USES: see below
CARE: None
PROPAGATION BY: Seedlings and cuttings
NATIVE OF: see below
OTHER: *S. americana*; e. United States; supporting stalk for a single flower; 1/2″ long or less; leaves mostly glabrous beneath; stream banks.
S. grandifolia; e. United States; supporting stalk for a single flower; 1/2″ long or less; leaves densely pubescent beneath; rich woods.
S. japonica; Japan, China; supporting stalk for a single flower 1″ long or more; rich moist soil.

Stranvaesia

Stranvaesia davidiana

Rose Family

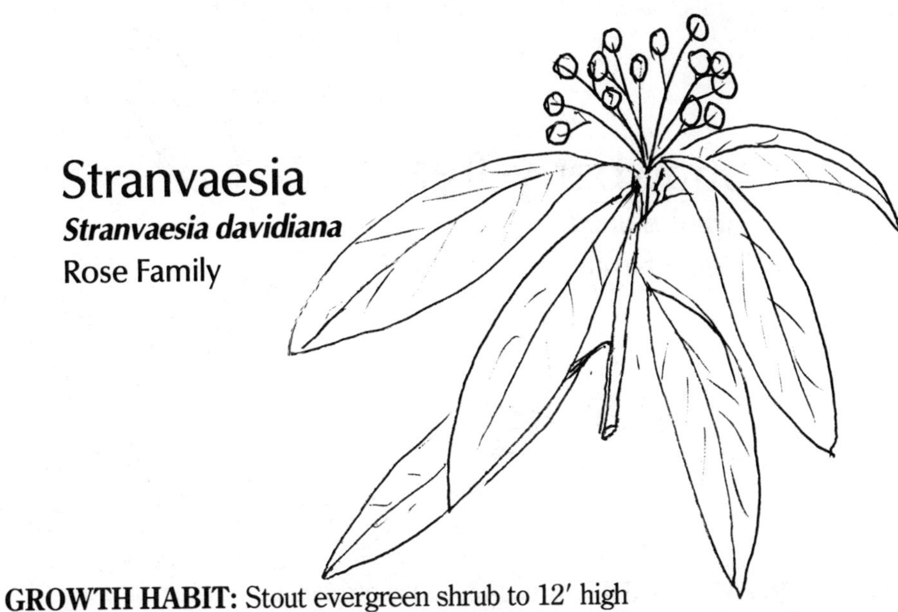

GROWTH HABIT: Stout evergreen shrub to 12′ high
LEAVES: Alternate, leathery and to 4″ long; leaf stalk present, teeth none
FLOWERS: White, hardly 1/2″ wide, and in 2-3″ wide loose terminal clusters
FRUITS: Red, berry-like and about 1/3″ wide
SEEDS: Few per fruit
LIGHT: Half sun or more
SOIL: Sandy loam
MOISTURE: Dry
SEASONAL ASPECT: Early summer flowers
ZONE: 1, 2, 3
USES: Screen, informal shrub border
CARE: Prune for size
PROPAGATION BY: Seeds, cuttings
NATIVE OF: w. China

Strawberry-Tree

Arbutus unedo

Heath family

GROWTH HABIT: Broad-topped evergreen shrub or small tree
LEAVES: Alternate, elliptic, toothed, to 3″ long
FLOWERS: Small, white, vase-shaped, about 1/4″ long and in short drooping racemes
FRUITS: 3/4″ raspberry-looking berries
SEEDS: Several and small
LIGHT: Sun or partial shade
SOIL: Rich loam
MOISTURE: Moist or dry
SEASONAL ASPECT: None
ZONE: 1, 2
USES: Specimen, occasional shrub in shrub border
CARE: None
PROPAGATION BY: Seeds or potted seedlings
NATIVE OF: United States west coast

Smooth Sumac

Rhus glabra

Cashew Family

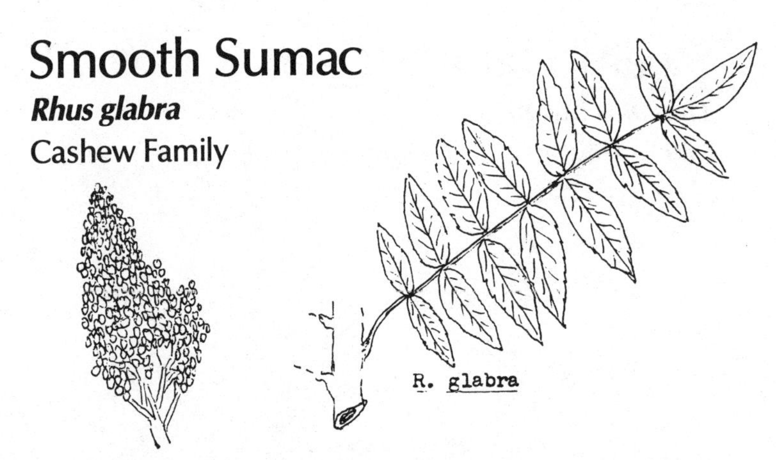

GROWTH HABIT: Deciduous shrub with large, unbranched stem; slowly colony forming

LEAVES: Alternate, pinnate with 11-31 lanceolate toothed leaflets 2-4″ long and whitened beneath

FLOWERS: Small, greenish and in dense terminal panicles

FRUITS: 1/8″ wide or less, red and covered with glandular hairs

SEEDS: 1

LIGHT: Full or partial sun

SOIL: Very tolerant

MOISTURE: Moist

SEASONAL ASPECT: Large, terminal panicles of red fruit and colorful autumn foliage

ZONE: 1, 2, 3

USES: Margin

CARE: None

PROPAGATION BY: Seedlings or sprouts

NATIVE OF: e. United States

OTHER: Fresh red berries and honey stirred together in cold water has been used as a substitute for "pink lemonade".

Winged Sumac; *R. copallina*; e. United States, generally smaller, more pubescent and with rachis or main vein of leaf (between the leaflets) with very narrow leaf-like wing; use in thin dry soil.

Fragrant Sumac; *R. aromatica*; e. United States; diffuse deciduous shrub to 4″ high; leaflets 3 and coarsely toothed; fruits red.

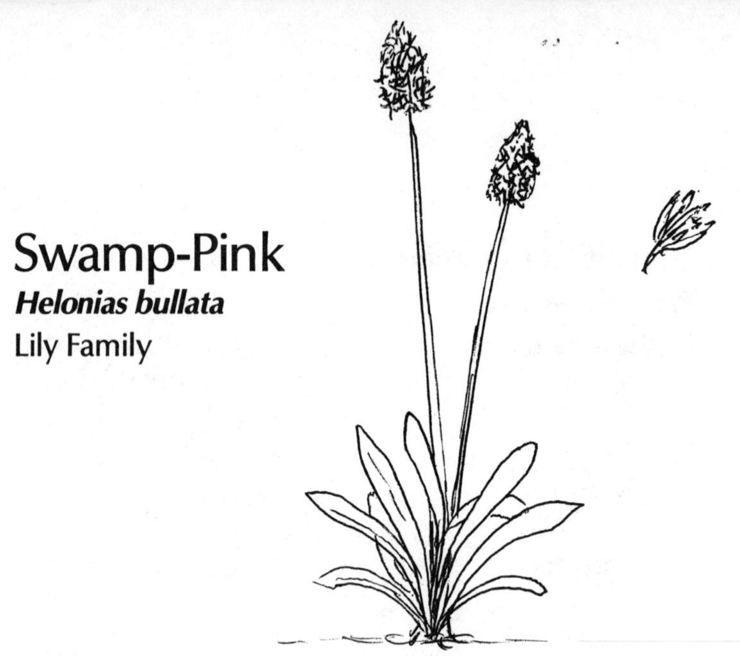

Swamp-Pink

Helonias bullata

Lily Family

GROWTH HABIT: Evergreen basal rosette former from stout rhizome
LEAVES: To 1 1/2″ long, wider toward tips, thickish and several in a tuft
FLOWERS: Pink, fragrant, about 1/2″ long and raised in a dense cluster above the leaves
FRUITS: 3-celled papery capsule
SEEDS: Narrow
LIGHT: Much shade
SOIL: Bog type
MOISTURE: Wet
SEASONAL ASPECT: Spring flowers
ZONE: 3
USES: Interest
CARE: None
PROPAGATION BY: Seeds
NATIVE OF: Mountains, s.e. United States, uncommon in the wild

Swamp Privet

Forestiera acuminata

Olive Family

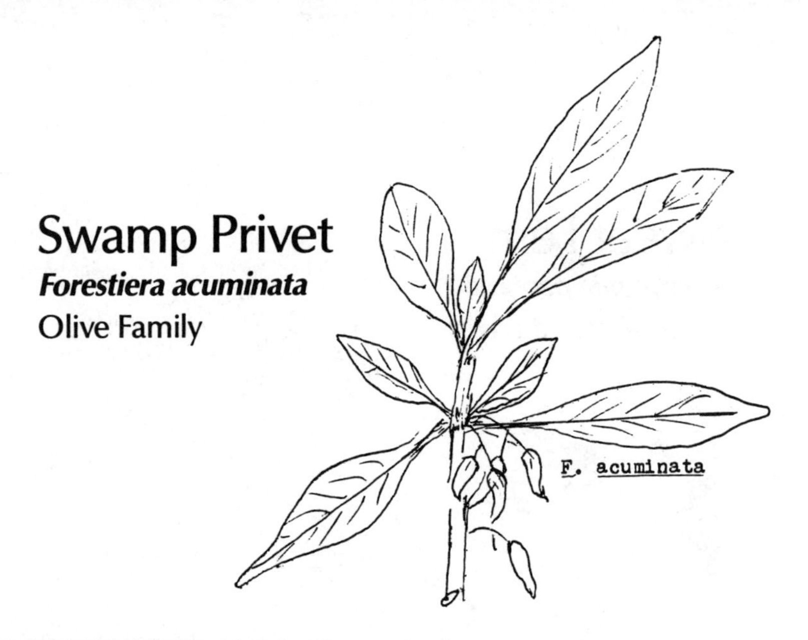

GROWTH HABIT: Tall deciduous shrub
LEAVES: Opposite, ovate, long pointed and to 4″ long
FLOWERS: Very small, no petals, green and in small clusters
FRUITS: 1/2″ long, black, elongated and curved
SEEDS: 1
LIGHT: Canopy shade
SOIL: Alluvial
MOISTURE: Wet
SEASONAL ASPECT: None
ZONE: 1, 2
USES: Pond or stream margin
CARE: None
PROPAGATION BY: Seeds
NATIVE OF: s.e. United States
OTHER: *F. ligustrina*; s.e. United States; smaller and less pointed leaves and smaller fruits; lives in moist to dry habitats.

Sweet Fern

Comptonia peregrina

Sweet-Gale Family

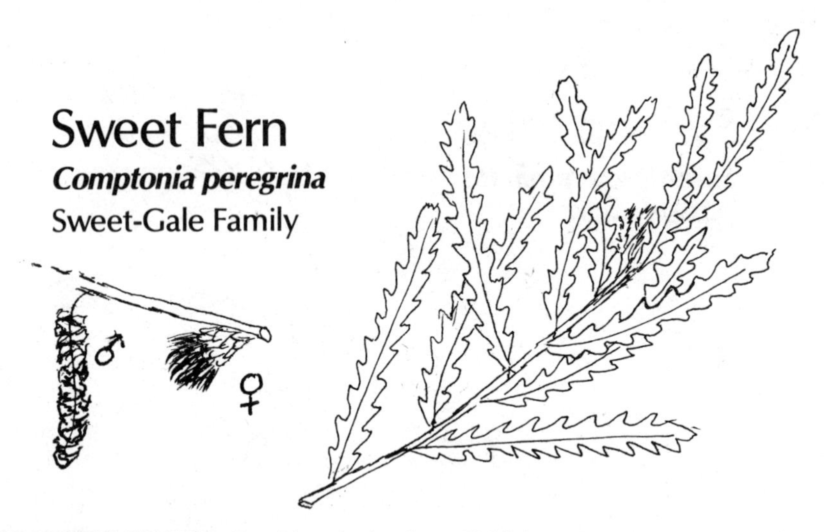

GROWTH HABIT: Deciduous shrub to 3′ high
LEAVES: Alternate, pinnately lobed to the point of being fern like, fragrant and to 4″ long
FLOWERS: Male as slender catkins; female catkin short
FRUITS: Female catkin becoming brown and bur-like
SEEDS: Several
LIGHT: Partial shade
SOIL: Usually thin
MOISTURE: Moist
SEASONAL ASPECT: None
ZONE: 2, 3
USES: To present fern-like appearance
CARE: None
PROPAGATION BY: Seeds or seedlings; transplanting difficult
NATIVE OF: e. United States
OTHER: Sometimes referred to as *Myrica asplenifolia.*

Sweet-Gum

Liquidambar styraciflua

Witch-Hazel Family

GROWTH HABIT: Large, deciduous tree, the bark of which produces a fragrant resin

LEAVES: Alternate, star-shaped with finely toothed margins, smooth and fragrant when crushed

FLOWERS: Male as condensed, yellowish head-like racemes produced before the leave; females as long peduncled balls of 2-beaked pistils

FRUITS: Round spiny balls of capsules 1″ wide

SEEDS: Fertile, dark and winged, and abortive as small saw dust, usually 1 good, 3 bad per capsule

LIGHT: Tolerant

SOIL: Tolerant

MOISTURE: Tolerant

SEASONAL ASPECT: Autumn color

ZONE: 1, 2, 3

USES: Background, shade, freestanding

CARE: None

PROPAGATION BY: Seedings

NATIVE OF: e. United States

OTHER: A very hardy, rapid-growing tree producing some late winter litter with the dry fruit balls. Gum formerly collected for medicinal use.

Sweet Pepper-Bush

Clethra spp.

White-Alder Family

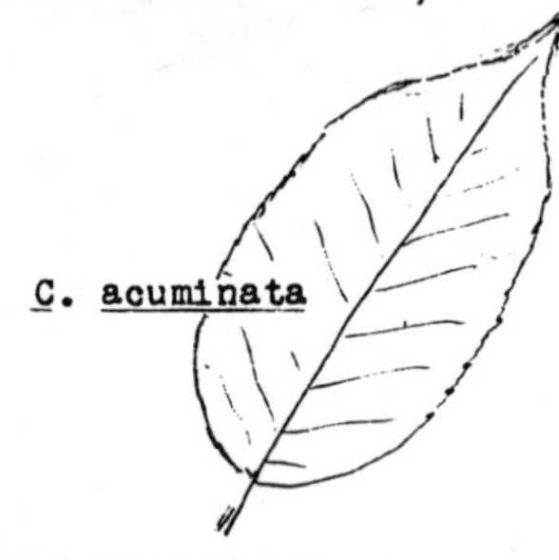

GROWTH HABIT: Deciduous shrub
LEAVES: Alternate, obovate to elliptic, finely toothed and 2-6″ long
FLOWERS: White, 1/2″ wide, fragrant and in terminal racemes
FRUITS: Very small pubescent capsules
SEEDS: Pinkish
LIGHT: Half shade, or more
SOIL: See below
MOISTURE: Moist
SEASONAL ASPECT: Summer flowers
ZONE: See below
USES: Blender, margins, specimen
CARE: None
PROPAGATION BY: Seedlings or sprouts
NATIVE OF: e. United States
OTHER: White Alder, *C. acuminata*; mountains; to 10′ high in rich moist place.

Sweet Pepper Bush; *C. alnifolia*; coastal plain; to 6′ high in moist, sandy places; colony forming.

Cinnamon Clethra; *C. barbinervis*; Japan; large shrub or small tree; obovate, toothed leaves; fragrant flowers.

Sweet-Shrub, Carolina Allspice

Calycanthus floridus

Calycanthus Family

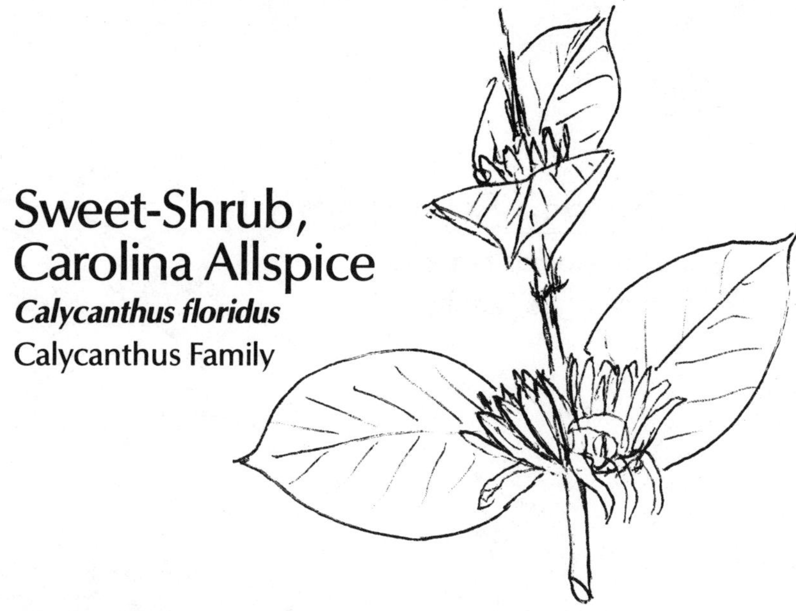

GROWTH HABIT: Deciduous, aromatic shrub to 8′ high
LEAVES: Opposite, entire, ovate to elliptic, 2-5″ long
FLOWERS: Before the leaves, brownish-purple, fragrant sepals and petals undifferentiated (flowers persist until leaves are grown).
FRUITS: A sort of sac 2-3″ long containing several nutlets
SEEDS: Up to 1/2″ long, brown
LIGHT: Full or partial shade
SOIL: Fertile loam, lots of humus
MOISTURE: Moist
SEASONAL ASPECT: Flower fragrance in early spring
ZONE: 1, 2, 3
USES: Occasional shrub, informal shrub border
CARE: None
PROPAGATION BY: Seeds or root sprouts
NATIVE OF: s.e. United States
OTHER: Some plants are more fragrant than others. Seeds toxic.

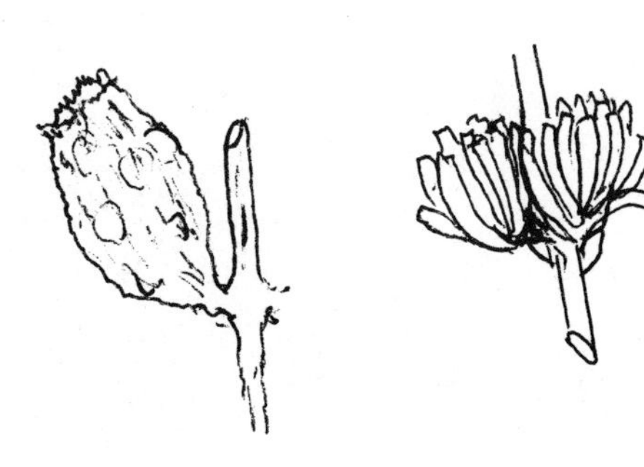

Chinese Sweet-Shrub

Chimonanthus praecox

Calycanthus Family

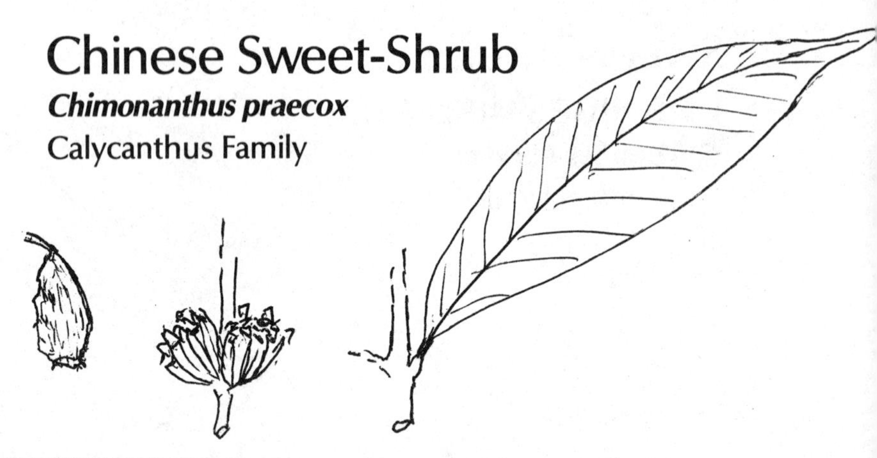

GROWTH HABIT: Deciduous shrub to 10′ high
LEAVES: Elliptic to ovate, entire and to 6″ long
FLOWERS: Yellow and appearing long before the leaves, not very fragrant but bees are attracted
FRUITS: A kind of sac 1-1 1/2″ long containing a few nutlets
SEEDS: Brown and to 1/4″ long
LIGHT: Sun or partial sun
SOIL: Fertile loam
MOISTURE: Moist
SEASONAL ASPECT: Yellow flowers in late winter
ZONE: 1, 2, 3
USES: Occasional shrub
CARE: None
PROPAGATION BY: Seeds or cuttings
NATIVE OF: China
OTHER: Seeds probably toxic.

Sycamore

Platanus occidentalis

Plane-Tree Family

GROWTH HABIT: Large, deciduous tree with spreading branches
LEAVES: Alternate, 4-10″ wide, shallowly 3-5 lobed and toothed, pubescent beneath
FLOWERS: Male in pendulous balls 1/2″ wide; female heads also ball-like, 1 1/2″ wide and many flowered
FRUITS: Achenes with tawny silky hairs
SEEDS: 1
LIGHT: Half sun or more
SOIL: Fertile silt or fine textured loam
MOISTURE: Moist
SEASONAL ASPECT: Conspicuous in winter because of the light colored bark of young trunks and branches mottled with patches of gray and green
ZONE: 1, 2, 3
USES: Freestanding, shade, street
CARE: None
PROPAGATION BY: Seedlings
NATIVE OF: e. United States
OTHER: A hardy tree, once established.

French Tamarisk

Tamarix gallica

Tamarisk Family

GROWTH HABIT: Semievergreen shrub or small tree
LEAVES: Alternate, sessile, gray-green, scale-like on slender flexuous branches
FLOWERS: Regular; 5-parted; white or pink and in slender 1-2″ long raceme-like arrangements; petals wither but remain for weeks
FRUITS: Very small capsule
SEEDS: Each bears a tuft of fine hair
LIGHT: Full sun
SOIL: Tolerant
MOISTURE: Moist
SEASONAL ASPECT: Late spring flowers
ZONE: 1 (2)
USES: Coastal planting, sandy area
CARE: None
PROPAGATION BY: Seedlings
NATIVE OF: w. Europe to the Himalayas
OTHER: Widely escaped along the Carolina coast. Four or more other species, all generally similar, are available in the trade. Salt tolerant.

Tar-Flower

Befaria racemosa

Heath Family

GROWTH HABIT: Evergreen shrub to 6′
LEAVES: Elliptic to oval and to 2″ long
FLOWERS: White to pink, to 2 1/2″ wide, 7-8 parted, and in narrow racemes
FRUITS: Depressed capsule
SEEDS: Several and very small
LIGHT: Partial shade underneath high canopy
SOIL: Sandy loam
MOISTURE: Moist to dry
SEASONAL ASPECT: Spring flowers
ZONE: 1
USES: Occasional flowering shrub, specimen, interest
CARE: None
PROPAGATION BY: Seedlings and cuttings
NATIVE OF: Lower coastal plain, Ga. - Miss.

Tea

Thea sinensis

Tea Family

GROWTH HABIT: Evergreen shrub
LEAVES: Alternate, toothed, glossy, 2-4″ long
FLOWERS: White, fragrant, axillary, 1 1/2″ wide
FRUITS: Large leathery capsule
SEEDS: 3-5 and glossy brown
LIGHT: Full or at least half sun
SOIL: Fertile loam
MOISTURE: Moist, but well drained
SEASONAL ASPECT: Flowering time in late spring
ZONE: 1, 2
USES: Specimen or interest planting
CARE: Similar to Camellia
PROPAGATION BY: Seeds, seedlings or cuttings
NATIVE OF: China
OTHER: One or more experimental tea plantations exist in South Carolina.

Tea Olive

Osmanthus spp.

Olive Family

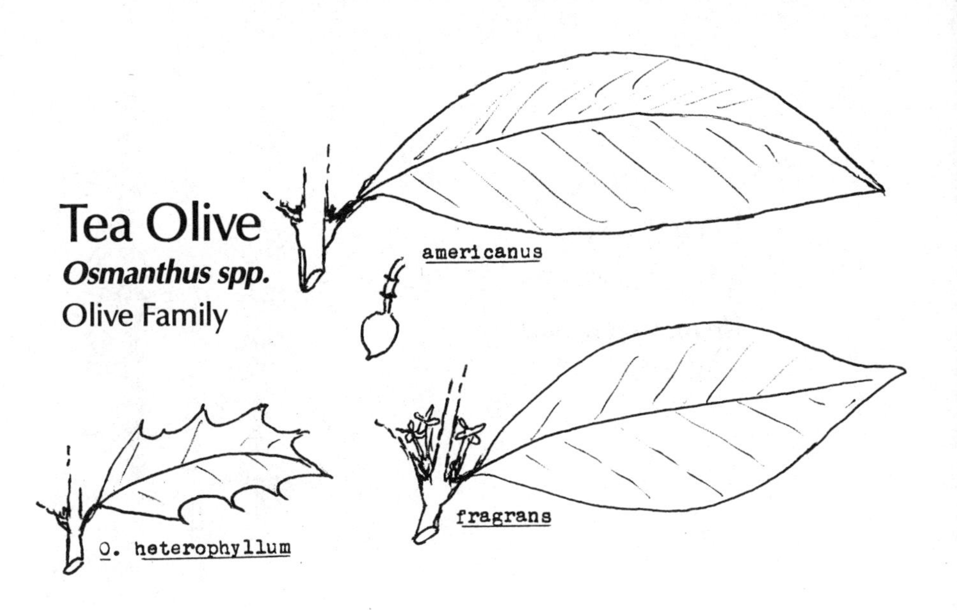

GROWTH HABIT: Large evergreen shrubs to 15′ or more

LEAVES: Opposite, somewhat clustered toward twig tips, entire, toothed or spiny

FLOWERS: Small, white, 4-parted, usually imperfect and very fragrant

FRUITS: Blue drupe about 1/2″ long with thin pulp

SEEDS: 1

LIGHT: Full sun or high, shifting shade

SOIL: Tolerant

MOISTURE: Moist to dry

SEASONAL ASPECT: The delightful fragrance from the spring and summer flowers permeates the surroundings

ZONE: 1, 2

USES: Freestanding, specimen, informal shrub border, fall fragrance

CARE: Some pruning to maintain size and shape

PROPAGATION BY: Cuttings

NATIVE OF: See below

OTHER: Wild Olive, *O. americanus*; s.e. United States; will attain tree size; leaves entire, thick, glossy above and to 6″ long.

Tea Olive, *O. fragrans*; Himalayas, China, Japan; leaves elliptic, to 4″ long, entire or finely and sharply toothed and of a dry, leathery texture.

Tea Olive, *O. heterophyllus*; Japan (*O. ilicifolius, O. aquifolius*); leaves dry-leathery, spiny and to 2 1/2″ long; flowers in fall. The first two have given rise to a number of varieties and hybrids. *O. X fortunei*; a hardy cross between above two imports.

Thyme, Mother-of-Thyme

Thymus serpyllum

Mint Family

GROWTH HABIT: Semievergreen subshrub with creeping pubescent stems rooting at the nodes, forming 2-3″ thick mat
LEAVES: Very small, 1/4″ long, entire and pointed
FLOWERS: Very small, pink, lavender or white and clustered terminally
FRUITS: 4 nutlets per flower
SEEDS: 1 per nutlet
LIGHT: Partial sun
SOIL: Adapted to most types
MOISTURE: Dry to moist
SEASONAL ASPECT: None
ZONE: 2, 3
USES: Bed, bank, rockery, edging
CARE: None
PROPAGATION BY: Rooted stems
NATIVE OF: n. Europe
OTHER: Several varieties.

Ti-ti, Leatherwood

Cyrilla racemiflora

Cyrilla Family

GROWTH HABIT: Semievergreen shrub
LEAVES: Alternate but crowded toward twig tips, thick, smooth, entire, oblanceolate and to 3 1/2″ long
FLOWERS: Small, white and crowded in 2-4″ long racemes
FRUITS: Very small and indehiscent, drupe-like
SEEDS: 1-2
LIGHT: Sun or partial shade
SOIL: Silt or silty loam
MOISTURE: Wet, moist or dry
SEASONAL ASPECT: Flowering time in early summer, and in fall when leaves show some color
ZONE: 1, 2
USES: Margin, wet spot, occasional flowering shrub
CARE: None
PROPAGATION BY: Seedlings
NATIVE OF: s.e. United States

Torreya

Torreya nucifera

Yew Family

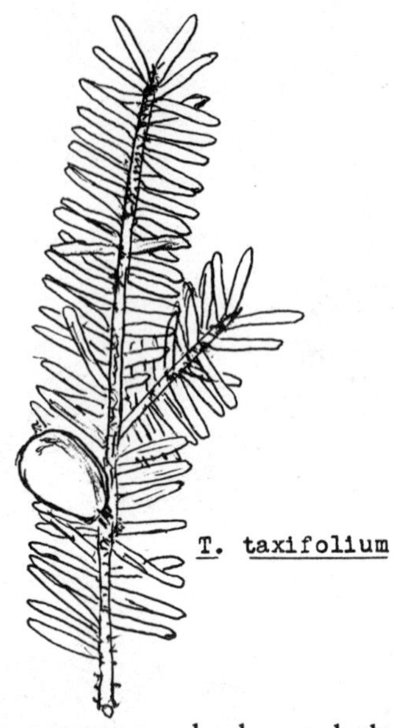

GROWTH HABIT: Large evergreen shrub, rarely becoming tree in cultivation; branches whorled
LEAVES: About 1″ long, lanceolate, 2-ranked, rigid and sharp pointed, glossy above
FLOWERS: Small, male in scaly heads; female solitary
FRUITS: 1″ long, purplish and with thin flesh
SEEDS: 1
LIGHT: Lots of sun
SOIL: Any porous fertile type
MOISTURE: Moist
SEASONAL ASPECT: None
ZONE: 1, 2
USES: Specimen
CARE: None
PROPAGATION BY: Seedlings
NATIVE OF: Japan
OTHER: *T. taxifolium*, medium size tree of the Apalachicola River basin of northwest Florida and south Georgia.

Trailing Arbutus

Epigaea repens

Heath Family

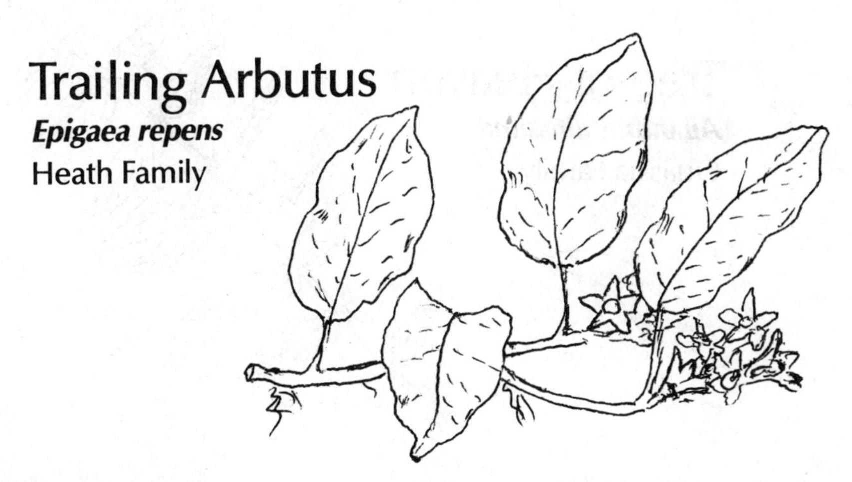

GROWTH HABIT: Prostrate, branching, evergreen shrub; patch forming
LEAVES: Alternate, oval to orbicular, to 2 1/2″ long
FLOWERS: Pink or white, 1/2″ or more long, very fragrant
FRUITS: Light-colored capsules
SEEDS: Very small
LIGHT: Partial sun
SOIL: Acid, sandy or loam
MOISTURE: Moist
SEASONAL ASPECT: Very early spring flowers
ZONE: 1, 2, 3
USES: Ground cover, bed, interest
CARE: None
PROPAGATION BY: Colony division
NATIVE OF: e. United States

Tree-of-Heaven

Ailanthus altissima

Quassia Family

GROWTH HABIT: Medium-sized, rapid growing, springly branched, deciduous tree
LEAVES: Alternate, 1-3′ long, pinnate with from 13-25 leaflets and each with 2-3 coarse teeth and foul smelling gland toward base
FLOWERS: Male and female flowers usually on separate trees
FRUITS: 2″ long, twisted, winged and in clusters
SEEDS: 1 per fruit
LIGHT: Full sun or less
SOIL: Any
MOISTURE: Very tolerant
SEASONAL ASPECT: Female trees in late summer when fruit is ripe
ZONE: 1, 2, 3
USES: Single specimen or where nothing else is likely to succeed
CARE: None
PROPAGATION BY: Seeds and root sprouts
NATIVE OF: China
OTHER: If in a suitable location this species spreads perniciously by root sprouts. Female trees have less leaf odor and, of course, no pollen. The tree of which it was said "A tree grows in Brooklyn".

Trumpet-Vine, Cow-Itch

Campsis radicans

Trumpet Creeper Family

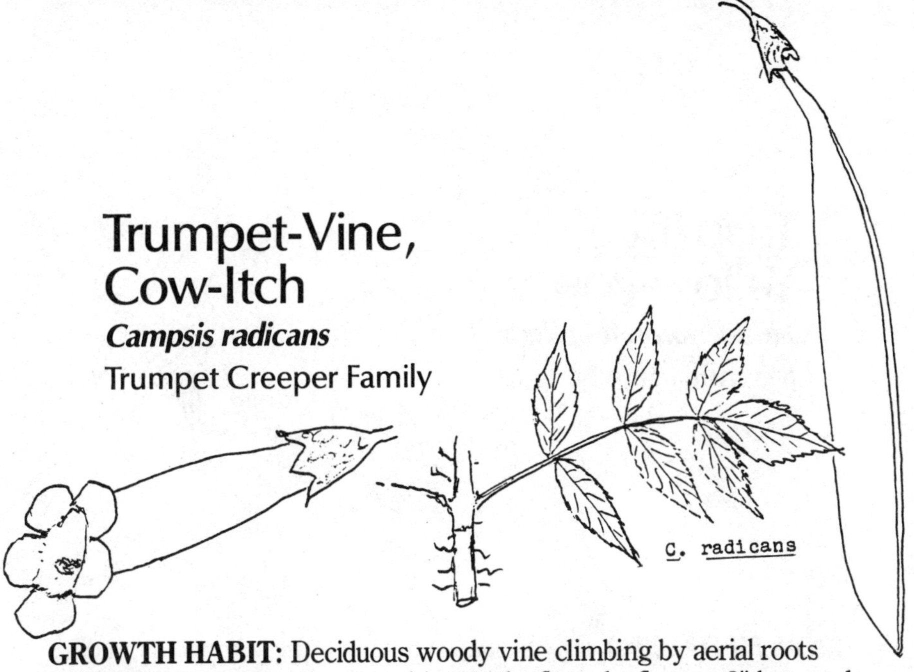

GROWTH HABIT: Deciduous woody vine climbing by aerial roots
LEAVES: Opposite, pinnate with 5-11 leaflets, leaflets to 2″ long and coarsely toothed
FLOWERS: Dull red or orange (one variety yellow) to 3″ long and 2″ wide with open center.
FRUITS: Woody pod to 8″ long
SEEDS: Many and winged
LIGHT: Full or partial sun
SOIL: Tolerant
MOISTURE: Moist or dry
SEASONAL ASPECT: Summer flowers
ZONE: 1, 2, 3
USES: Screen, fence, arbor
CARE: None
PROPAGATION BY: Seedlings or seeds
NATIVE OF: e. United States
OTHER: *C. grandiflora*; China; larger leaves and flowers.

Tulip-Tree, Yellow-Poplar

Liriodendron tulipifera

Magnolia Family

GROWTH HABIT: Large deciduous tree with single trunk, ascending branches and light gray bark that is smooth when young
LEAVES: Alternate, smooth, long petiole, broadly notched at apex and with 1 or 2 shallow lateral lobes
FLOWERS: Solitary at branch tips, greenish-yellow with orange and 1½″ long, borne with the new leaves
FRUITS: Many tightly spiraled on elongate receptacle
SEEDS: Angled and conspicuously winged
LIGHT: Some shade
SOIL: Fertile
MOISTURE: Moist
SEASONAL ASPECT: Yellow leaves in autumn
ZONE: 1, 2, 3
USES: Freestanding, border, stream bank
CARE: None
PROPAGATION BY: Seedlings
NATIVE OF: e. North America
OTHER: Rapid growing and important timber tree

Tung-Oil-Tree

Aleurites fordii

Spurge Family

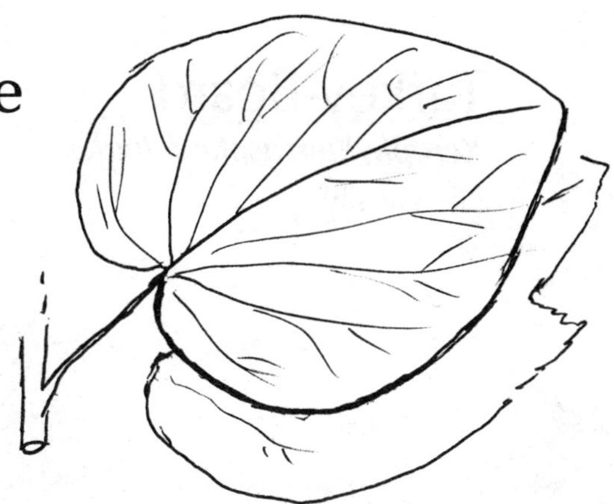

GROWTH HABIT: Small, spreading, deciduous tree with milky juice
LEAVES: Alternate, ovate, sometimes 3-lobed, 3-5″ long
FLOWERS: Male and female separate but on same tree, reddish-white about 2″ wide and before the leaves
FRUITS: Globose or somewhat flattened 2 1/2″ across and indehiscent
SEEDS: 2-3 rough surfaced nuts
LIGHT: Full sun
SOIL: Most any good soil
MOISTURE: Moist
SEASONAL ASPECT: Flowers, before the leaves
ZONE: 1, 2
USES: Specimen, shade
CARE: None
PROPAGATION BY: Seeds
NATIVE OF: Central Asia
OTHER: Seeds are a source of oil and are toxic.

Turkey-Beard

Xerophyllum asphodeloides

Lily Family

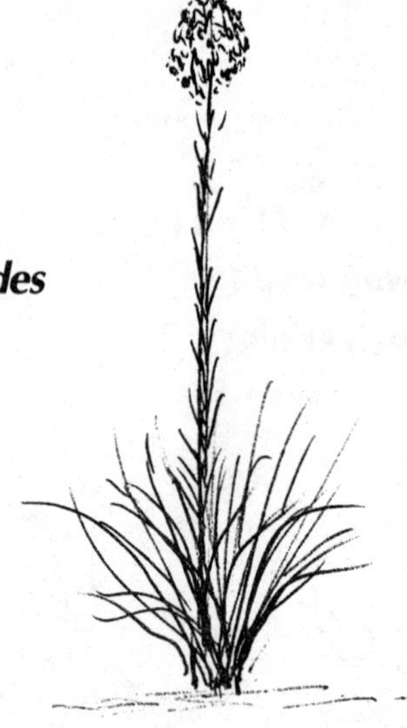

GROWTH HABIT: Stout, nearly evergreen rosette-former from root crown
LEAVES: Narrowly linear, to 2 1/2′ long, forming dense tuft
FLOWERS: White, about 1/4″ long, numerous, borne on condensed panicle raised to twice the height of the basal leaves
FRUITS: 3-celled capsule
SEEDS: Narrow and glossy brown
LIGHT: Partial sun
SOIL: Fertile loam
MOISTURE: Moist to dry
SEASONAL ASPECT: Flowers in late spring
ZONE: 3
USES: Specimen plant
CARE: None
PROPAGATION BY: Seeds or seedlings
NATIVE OF: Mountains, s.e. United States

Turk's Fez

Malvaviscus arboreus

Mallow Family

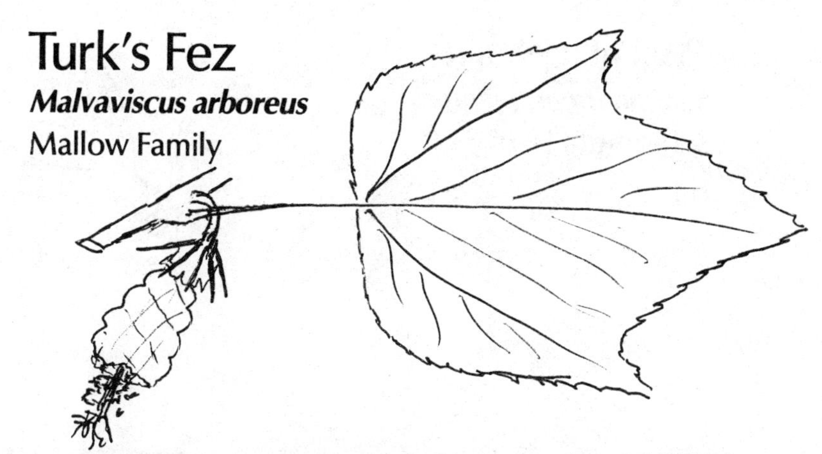

GROWTH HABIT: Low, pubescent, deciduous shrub to 2 1/2′ high
LEAVES: Alternate, velvety to the touch, ovate, to 4″ long, margin toothed and sometimes angled
FLOWERS: Scarlet, 1 1/2″ long, open but not spreading, with narrow bracts below the calyx
FRUITS: Somewhat berry-like body of fleshy carpels
SEEDS: Several
LIGHT: Partial or full sun
SOIL: Fertile
MOISTURE: Moist
SEASONAL ASPECT: Summer flowers
ZONE: 1, 2
USES: Accent, decorations, interest, occasional flowering shrub
CARE: None
PROPAGATION BY: Seedlings
NATIVE OF: Mexico and South America

Twining-Vine

Schisandra propinqua

Schisandra Family

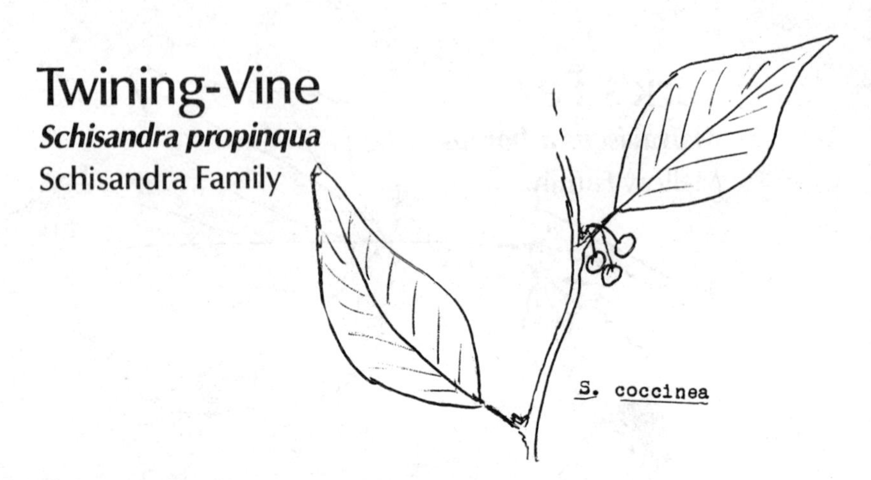

GROWTH HABIT: Evergreen woody vine climbing by twining
LEAVES: Alternate, broadly elliptic, pointed and to 4″ long
FLOWERS: Orange, 3/4″ wide and long stalked from leaf axis; male and female on separate plants
FRUITS: Red berries
SEEDS: Few per berry
LIGHT: Sun or partial shade
SOIL: Sandy
MOISTURE: Moist
SEASONAL ASPECT: Red fruits in fall
ZONE: 1
USES: Arbor, trellis, specimen
CARE: Trimming for shape
PROPAGATION BY: Seedlings
NATIVE OF: Himalayas
OTHER: *S. coccinea (glabra)*; s.e. United States; a very rare native species.

Viburnum

Viburnum spp.

Honeysuckle Family

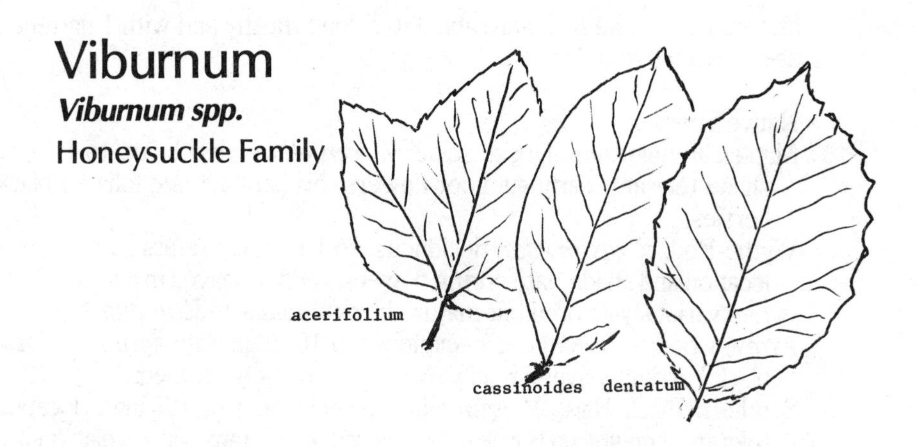

The Viburnums represent a large and popular group of ornamental shrubs with over 100 species, varieties and hybrids in the trade. One of their values is that supposedly they somehow enhance most kinds of nearby shrubbery. Some are native; most are introduced. All have opposite leaves. Some species are evergreen, others deciduous. The texture of the foliage varies greatly and some show vivid autumn colors. The attraction for some is that they produce showy flower masses in the spring, while for others it is the bright colored fruits in late summer and fall. Individual flowers are small, white, 5-parted and are borne in flat, rounded or pyramidal clusters. In some the peripheral flowers are sterile

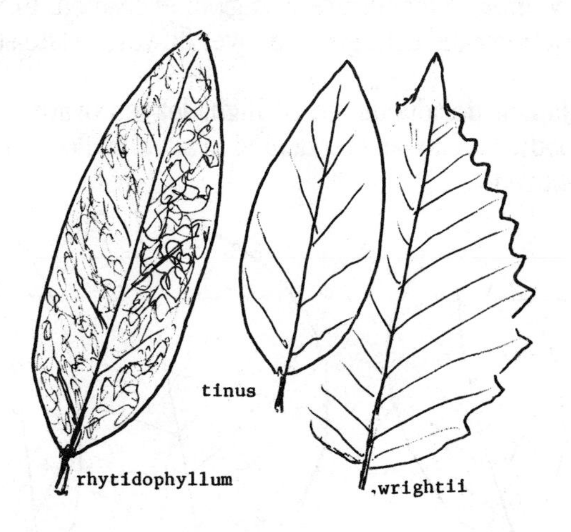

and enlarged. The fruits are about 1/2″ long, fleshy and with 1 flattened seed. Attractive to birds.

Native Species

Mapleleaf Vibiurnum, Dockmackie; *V. acerifolium*; deciduous; to 4′ high; shade tolerant, early summer flowers; brilliant autumn foliage; black berries.

Wythe-Rod; *V. cassinoides*; deciduous; to 10′ high; prefers moist to wet location and shade; late spring flowers; reddish leaves in autumn; berries variously colored but mature black. Similar to *V. nudum.*

Arrow-wood; *V. dentatum*; deciduous, to 10′ high; late spring flowers, black berries, leaves broadly ovate and coarsely toothed.

Southern Black Haw; *V. prunifolium*; deciduous; to 12′ high; location tolerant; late spring flowers; leaves red in autumn; berries black; similar to V. rufidulum.

Introduced Species

Chinese Snowball; *V. macrocephalum*; semievergreen; to 10′ high; leaves oval, toothed and to 4″ long; flowers in mid-spring.

Sweet Viburnum; *V. odoratissimum*; India, Japan; evergreen; to 10′ high; leaves thick, elliptic, glossy above and to 6″ long; mid-spring flowers; fruits black.

Leatherleaf Viburnum; *V. rhytidophyllum*; China; evergreen; to 8′ high; leaves ovate, mostly entire, to 6″ long, deeply wrinkled, lustrous above, soft-hairy beneath; flowers in late summer; fruits red, becoming black.

Laurestinus; *V. tinus*; Mediterranean region; evergreen; to 8′ high; leaves elliptic, to 3″ long and dark green above; flowers in late summer; fruits black.

V. wrightii; Japan; deciduous; to 8′ high; leaves ovate, to 6″ long and coarsely toothed, veiny and reddening in autumn; flowers in late spring; fruits bright red.

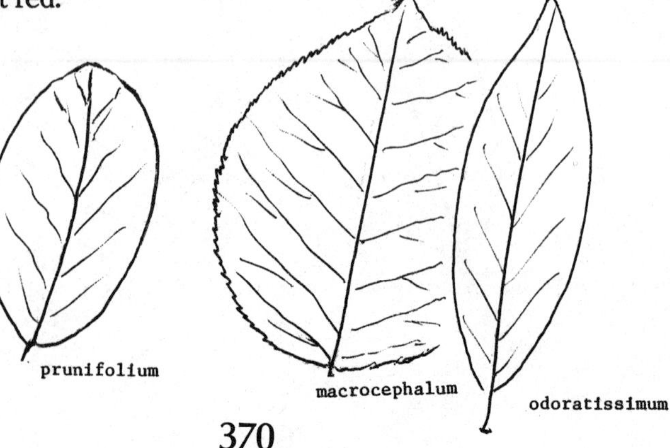

Virginia Willow

Itea virginica

Saxifrage Family

GROWTH HABIT: Deciduous shrub to 6′ high, with slender green stems

LEAVES: Alternate, elliptic, finely toothed and to 4″ long

FLOWERS: Small, white, 5-parted and in narrow terminal racemes

FRUITS: Small 2-celled capsules

SEEDS: Tiny

LIGHT: Half to full shade

SOIL: Sandy clay

MOISTURE: Wet

SEASONAL ASPECT: Late spring flowers

ZONE: 1, 2, 3

USES: Specimen

CARE: None

PROPAGATION BY: Root sprouts, cuttings

NATIVE OF: s. e. United States

Black Walnut

Juglans nigra

Walnut Family

GROWTH HABIT: Large spreading deciduous tree much valued for shade, nuts and lumber
LEAVES: Alternate, odd pinnate with 13-21 aromatic leaflets
FLOWERS: Male in drooping catkins; female inconspicuous
FRUITS: Rough-surfaced nut inside thick indehiscent husk
SEEDS: Large and edible
LIGHT: Plenty of sun
SOIL: Rich loam
MOISTURE: Moist
SEASONAL ASPECT: None
ZONE: 1, 2, 3
USES: Many
CARE: None
PROPAGATION BY: Seeds or seedlings
NATIVE OF: e. North America
OTHER: Walnut lumber is probably the most valuable wood in the United States. Butternut, White Walnut, *J. cinerea*; fruits obviously longer than wide; mountain area.

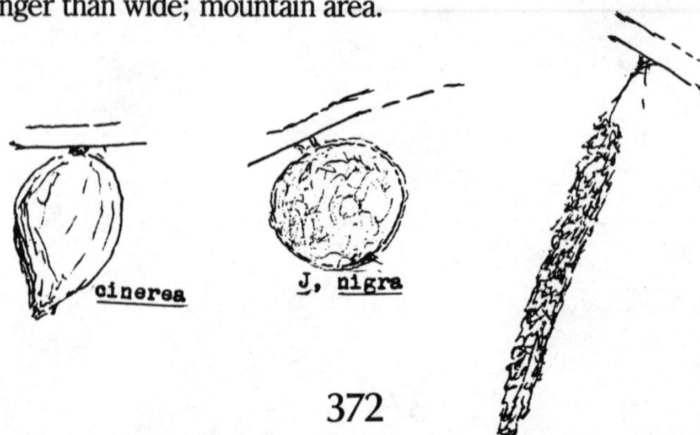

Water Elm

Planera aquatica

Elm Family

GROWTH HABIT: Small spreading deciduous tree
LEAVES: Alternate, ovate, toothed and to 2″ long
FLOWERS: Small, scaly, separate but on same tree
FRUITS: About 1/4″ wide and covered with leathery spines
SEEDS: 1
LIGHT: Partial sun
SOIL: Silty loam
MOISTURE: Moist or wet, until established
SEASONAL ASPECT: None
ZONE: 1, 2
USES: Interest, specimen, wet margin
CARE: None
PROPAGATION BY: Seeds or seedlings
NATIVE OF: s.e. United States

Water Willow

Decodon verticillatus

Loosestrife Family

GROWTH HABIT: Shrubby, colony-forming perennial
LEAVES: Opposite or whorled; lanceolate, entire and to 8′ or more long
FLOWERS: Pink or purple, 1/2″ long and in axillary clusters
FRUITS: Dark brown capsules
SEEDS: Reddish
LIGHT: Full or partial sun
SOIL: Any frequently inundated or shallowly flooded type
MOISTURE: Wet, until established
SEASONAL ASPECT: Summer flowers
ZONE: 1, 2
USES: Pond margin, shallow pond, marsh front
CARE: None
PROPAGATION BY: Rhizomes, seeds
NATIVE OF: s.e. United States

Wax-Myrtle, Bayberry

Myrica cerifera

Sweet-Gale Family

GROWTH HABIT: Large, nearly evergreen shrub to 12′
LEAVES: Alternate, aromatic, slightly toothed and rusty glandular-dotted beneath
FLOWERS: Inconspicuous with male and female on different plants
FRUITS: Grayish blue, wax coated and about 1/8′ wide
SEEDS: 1 per fruit
LIGHT: Full sun
SOIL: Sandy or silt loam
MOISTURE: Moist
SEASONAL ASPECT: None
ZONE: 1, 2
USES: Single planting, open border, side of pool or garden planting
CARE: None
PROPAGATION BY: Seedling
NATIVE OF: s.e. United States
OTHER: The thin wax covering on the berries may be removed in hot water and mixed with paraffin for candle making.

Weavers Broom

Spartium junceum

Legume Family

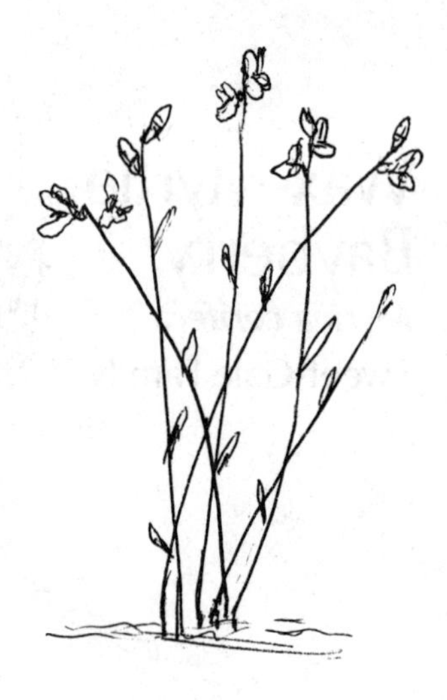

GROWTH HABIT: Deciduous, but almost leafless shrub to 6′ high; with green rush-like stems
LEAVES: Alternate, few, narrow, 1″ long
FLOWERS: Fragrant, yellow, 1′ long, calyx 1-lipped
FRUITS: Flat pod to 3″ long
SEEDS: Several
LIGHT: Full sun
SOIL: Most types
MOISTURE: Dry
SEASONAL ASPECT: Summer flowers, but only few
ZONE: 1, 2
USES: Hot, dry places
CARE: Transplant potted, prune regularly
PROPAGATION BY: Seedlngs or cuttings
NATIVE OF: Mediterranean region and the Canary Islands
OTHER: Somewhat similar to *Cytisus* and *Sophora*.

Weigela

Weigela florida

Honeysuckle Family

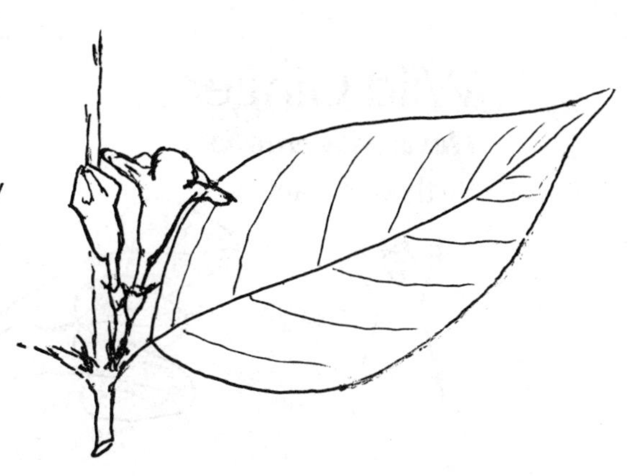

GROWTH HABIT: Decidous shrub to 8′ high, spreading by many branches arising from the base
LEAVES: Opposite, elliptic or wider, long-pointed and toothed.
FLOWERS: Pinkish, but showing some white and rose
FRUITS: Narrow woody capsule
SEEDS: Very small
LIGHT: Full or nearly full sun
SOIL: Tolerant to most types
MOISTURE: Moist
SEASONAL ASPECT: Late spring flowers
ZONE: 1, 2, 3
USES: Occasional flowering shrub
CARE: Some pruning yearly
PROPAGATION BY: Cuttings
NATIVE OF: China and Korea
OTHER: Hybrid varieties exist.

Wild Ginger

Hexastyllis arifolia

Birthwort Family

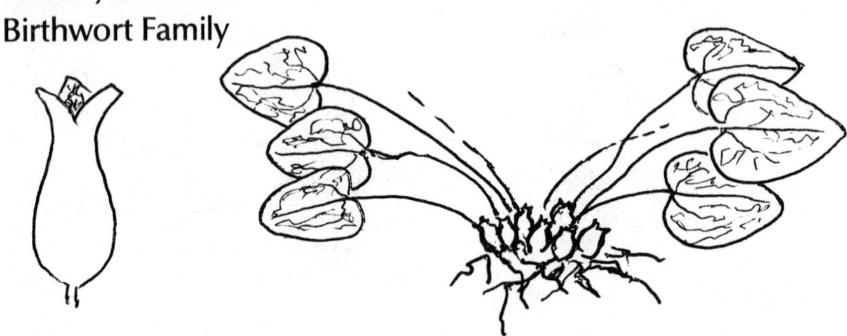

GROWTH HABIT: Low, stemless, fragrant evergreen
LEAVES: Reclining, long petioled, mottled and arrow- or heart-shaped
FLOWERS: 1″ long, flask-shaped, petalless and formed beneath the litter
FRUITS: Fleshy capsule
SEEDS: Several
LIGHT: Lots of shade
SOIL: Fertile with lots of humus
MOISTURE: Moist to dry
SEASONAL ASPECT: None
ZONE: 1, 2, 3
USES: Interest
CARE: None
PROPAGATION BY: Seeds or clump division
NATIVE OF: s.e. United States
OTHER: Formerly used as a medicinal or for a pleasant tea.

Willow

Salix spp.

Willow Family

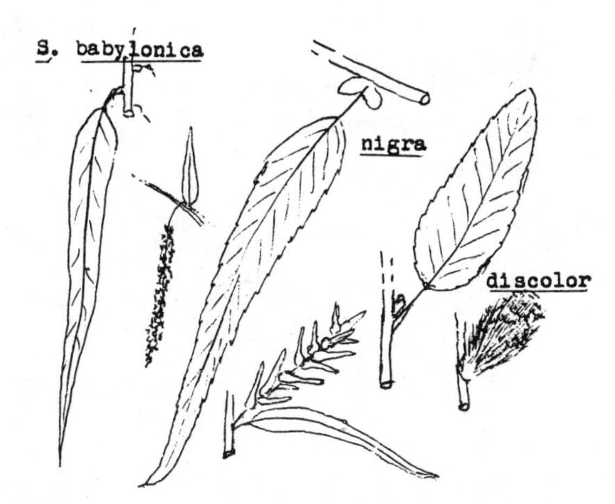

GROWTH HABIT: Deciduous trees and shrubs with lithe branches; many hybrids and selected clones

LEAVES: Alternate, mostly lanceolate or narrowly so, to elliptic?

FLOWERS: Male and female separate but both as catkins

FRUITS: Small capsules

SEEDS: Tiny and covered with long white silky hairs

LIGHT: Plenty of sun

SOIL: Heavy or fine textured

MOISTURE: Moist to wet

SEASONAL ASPECT: None

ZONE: 1, 2, 3

USES: Wet margin or bank, low place, informal shrub border

CARE: None

PROPAGATION BY: Cuttings

NATIVE OF: see below

OTHER: Weeping Willow, *S. babylonica,* China; lesser branches very long and pendant; several varieties.

Common Willow, S. *nigra,* e. United States; fast growing if in wet place.

Pussy Willow, S. *discolor,* n.e. North America; shrub with large silky buds in early spring.

Corkscrew Willow, S. *matasudana var. tortuosa*; s.e. Asia; to 30′ high. Once established Willows do well in most moist or even dry soils.

Winter Hazel

Corylopsis spp.

Witch Hazel family

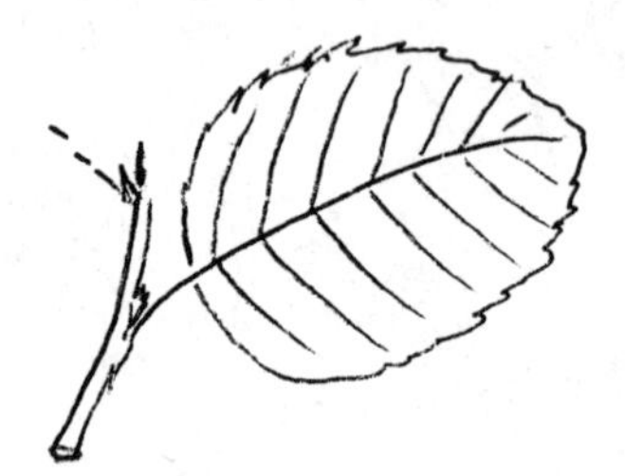

GROWTH HABIT: Deciduous shrubs to 8′ or more high
LEAVES: Alternate, prominently veined, coarsely toothed
FLOWERS: Yellow, 2-several in drooping arrangement each almost 1″ wide
FRUITS: Capsule
SEEDS: 2
LIGHT: Full sun
SOIL: Fertile type
MOISTURE: Moist
SEASONAL ASPECT: Spring flowers before the leaves
ZONE: 1, 2, 3
USES: Occasional shrub
CARE: None
PROPAGATION BY: Soft or semi-hardwood cutting, layering
NATIVE OF: Japan
OTHER: C. glabrescens; very fragrant.
C. *pauciflora*; flowers 2-3 together.
C. *spicata*; flowers several together in terminal spike; bush to only 3′ high.

Wire-Vine

Muehlenbeckia complexa

Smartweed Family

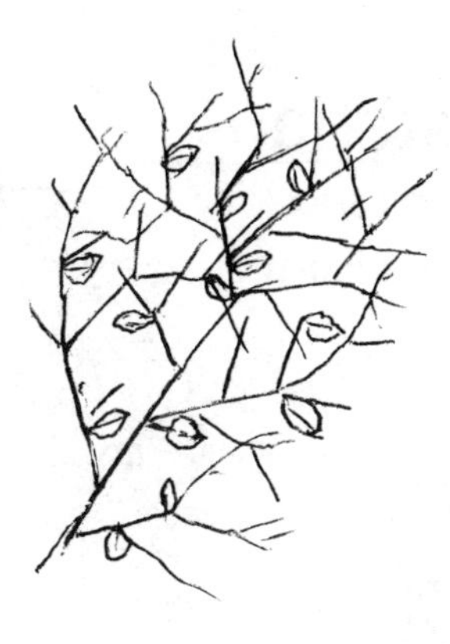

GROWTH HABIT: Strong evergreen climber or creeper with slender reddish wiry and much-branched stem
LEAVES: Alternate, circular to fiddle-shaped and to 1/2″ long
FLOWERS: Few, waxy-white and in terminal spike
FRUITS: 3-angled achene
SEEDS: 1
LIGHT: Full sun
SOIL: Sandy
MOISTURE: Dry
SEASONAL ASPECT: Summer growth
ZONE: 1
USES: Cover for stumps, rock, chimney, ground or hanging basket
CARE: Some trimming
PROPAGATION BY: Seeds
NATIVE OF: New Zealand

Wisteria

Wisteria spp.

Legume Family

GROWTH HABIT: Wildly climbing, twining, deciduous woody vine, becoming perhaps 6″ in diameter at base and reaching to the top of tall trees; a strangler

LEAVES: Alternate, pinnate, 7-19 lanceolate or wider leaflets 1-3″ long each

FLOWERS: 1″ long, violet or white and in long drooping racemes

FRUITS: Short, broad velvety or smooth pod

SEEDS: 1-2

LIGHT: Partial or full sun

SOIL: Tolerant

MOISTURE: Moist

SEASONAL ASPECT: Flowers in spring before the leaves

ZONE: 1, 2, 3

USES: Arbor, trellis

CARE: Pruning to keep it as shrub or to prevent it from strangling trees

PROPAGATION BY: Seedlings and layering

NATIVE OF: See below

OTHER: Chinese Wisteria, *W. sinensis*; 7-11 leaflets.

Japanese Wisteria, *W. floribunda*; 13-19 leaflets.

Native Wisteria, *W. frutescens*; glabrous pods with short racemes, a weak climber.

Witch-Alder

Fothergilla gardenii

Witch Hazel Family

GROWTH HABIT: Colony forming, deciduous shrub to 4′ high with stellate-pubescent twigs
LEAVES: Ovate with rounded tips, wavy or shallowly toothed margins and stellate-pubescence
FLOWERS: Small, white and in dense spikes, anthers raised on conspicuous white filaments
FRUITS: Densely hairy ovoid capsule
SEEDS: 1 per cell
LIGHT: Shade, or little sun
SOIL: Sandy or silty loam with peat
MOISTURE: Moist but well drained
SEASONAL ASPECT: Early spring flowers
NONE: 1, 2
USES: Interest
CARE: None
PROPAGATION BY: Colony division
NATIVE OF: s.e. United States
OTHER: A rare and larger form (*F. major*) occurs in the high Piedmont and mountains.

Witch-Hazel

Hamamelis virginiana

Witch Hazel Family

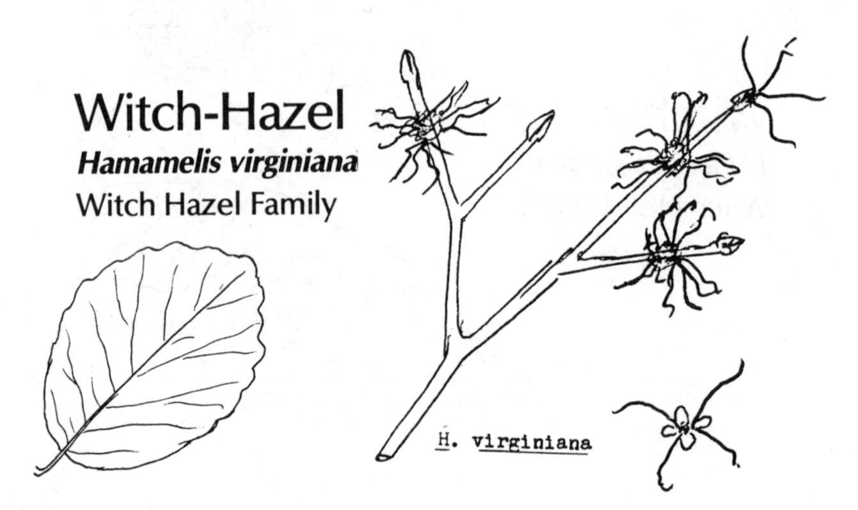

GROWTH HABIT: Deciduous shrub to 12′ high
LEAVES: Obovate, margin wavy or coarsely but shallowly toothed
FLOWERS: Yellow, petals 3/4″ long but very narrow, blooming in autumn after leaves fall
FRUITS: Woody capsule about 1/2″ long
SEEDS: Few
LIGHT: Shade or little sun
SOIL: Fertile
MOISTURE: Moist
SEASONAL ASPECT: Flowering time in autumn
ZONE: 1, 2, 3
USES: Occasional shrub, interest, specimen
CARE: None
PROPAGATION BY: Seedlings and cuttings
NATIVE OF: s.e. United States
OTHER: Two oriental species, generally similar to above, but with more and larger flowers are:
H. mollis; China; to 25′ high; leaves broadly obovate and 4-5″ long; flowers yellow and in late winter.
H. japonica; Japan; shrub, leaves 2-4″ long. Dowsers (Water Witchers) have long preferred to use Witch Hazel branches.

Woody Goldenrod

Solidago pauciflosculosa
(Chrysoma)

Composite Family

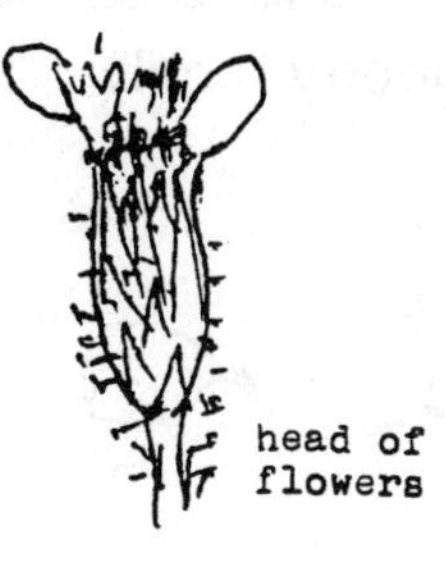

GROWTH HABIT: Freely branching evergreen shrub to 3′ high
LEAVES: Alternate, oblanceolate, round-tipped and with plated surface
FLOWERS: In 5-flowered heads, rays very short, bracts and branches of inflorescence yellow and sticky glandular
FRUITS: Very small achenes
SEEDS: One per achene
LIGHT: Full or lots of sun
SOIL: Very sandy
MOISTURE: Dry
SEASONAL ASPECT: Flowering time in late fall
ZONE: 1
USES: Specimen, interest
CARE: Keep down nearby competition
PROPAGATION BY: Seeds and seedlings
NATIVE OF: Sand hills, coastal plain, N.C. - Miss.

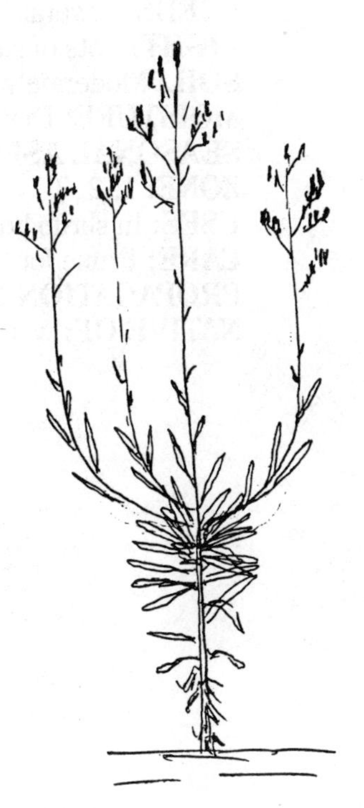

Xanthoceras

Xanthoceras sorbifolia

Soapberry Family

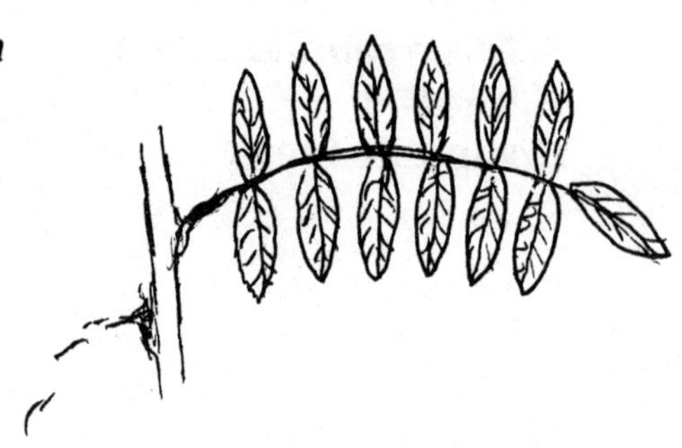

GROWTH HABIT: Deciduous shrub to 10; high
LEAVES: Alternate, odd-pinnate, to 1′ long, leaflets 9-17, each to 1 1/2″ long, without stalk and toothed
FLOWERS: White, 3/4″ wide and with brown or red spot on base of each petal
FRUITS: Thick-walled top-shaped capsule capsule to 2″ long
SEEDS: Several, 1/2″ long and brown
LIGHT: Lots of sun
SOIL: Moderately fertile loam
MOISTURE: Dry to moist
SEASONAL ASPECT: Flowers in mid-spring
ZONE: 1, 2, 3
USES: In shrub border, occasional flowering shrub
CARE: Prune for shape and size
PROPAGATION BY: Seeds, winter root cuttings
NATIVE OF: w. China

Yellow Archangel

Lamiastrum galeobdolon
(Lamium luteum)

Mint Family

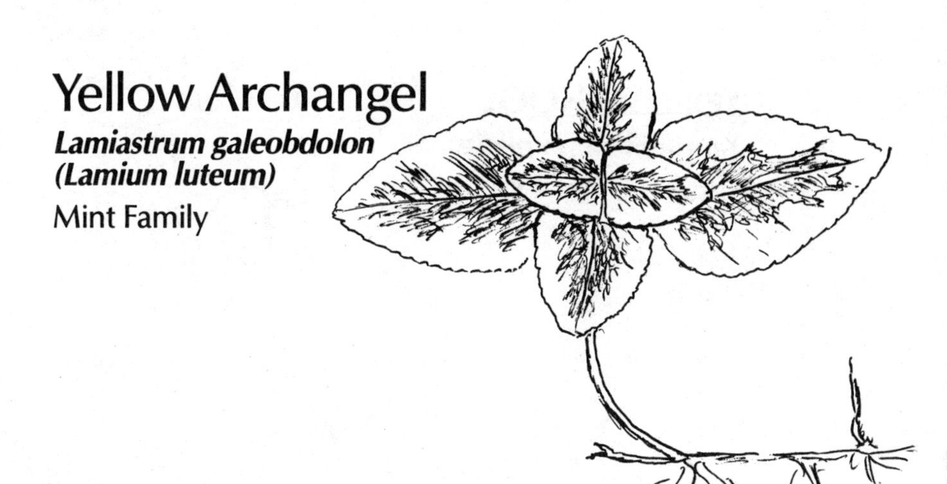

GROWTH HABIT: Evergreen, or nearly so, trailing perennial
LEAVES: Opposite, heart-shaped, round-toothed and usually variegated, to 2 1/2″ long
FLOWERS: Yellow, corolla tube longer than lobes, lip arched
FRUITS: Very small nutlets
SEEDS: 1 per nutlet
LIGHT: Mostly shade
SOIL: Fertile
MOISTURE: Moist
SEASONAL ASPECT: None
ZONE: 1, 2 (3)
USES: Ground cover, pot, hanging basket
CARE: None or maybe containment but not a vigorous spreader
PROPAGATION BY: Stolons
NATIVE OF: Europe
OTHER: The rather large leaves together with the conspicuous amount of white makes this plant a year around attraction.

Yellow-Root

Xanthorhiza simplicissima

Buttercup Family

GROWTH HABIT: Small deciduous shrub tending to form small colonies from the roots, which when broken are yellow; 1-2′ high

LEAVES: Long petioled and 1- or 2-pinnate; leaflets mostly 5, ovate and toothed to deeply lobed

FLOWERS: Small, brownish-purple and in racemes

FRUITS: Several very small thin-walled pods per flower

SEEDS: Very small

LIGHT: Shade

SOIL: Silty or sandy loam

MOISTURE: Moist to wet

SEASONAL ASPECT: None

ZONE: (1), 2, 3

USES: Interest

CARE: None

PROPAGATION BY: Suckers

NATIVE OF: e. United States

OTHER: An interesting addition to a shaded streambank. It has long been sought by herb collectors for its reputed medicinal value.

Yellow-Wood

Cladrastis lutea

Legume Family

GROWTH HABIT: A beautiful flowering, native deciduous tree with short trunk, spreading crown, smooth gray bark and yellow wood

LEAVES: Alternate, pinnate, 7-9 leaflets; each ovate, 3-4″ long and entire

FLOWERS: 1″ long, white, fragrant and in many-flowered drooping panicles 8-15″ long

FRUITS: 2-4″ long pods

SEEDS: Several

LIGHT: Sun or partial shade

SOIL: Loose and fertile

MOISTURE: Moist

SEASONAL ASPECT: Late spring flowers and bright yellow leaves in autumn

ZONE: 1, 2, 3

USES: Decoration, background, freestanding, interest

CARE: None

PROPAGATION BY: Seeds

NATIVE OF: Missouri east to the Carolinas

OTHER: *C. chinensis*; China; is a multi-stemmed shrub.

Zamia, Coontie, Florida Arrowroot

Zamia integrifolia

Cycad Family

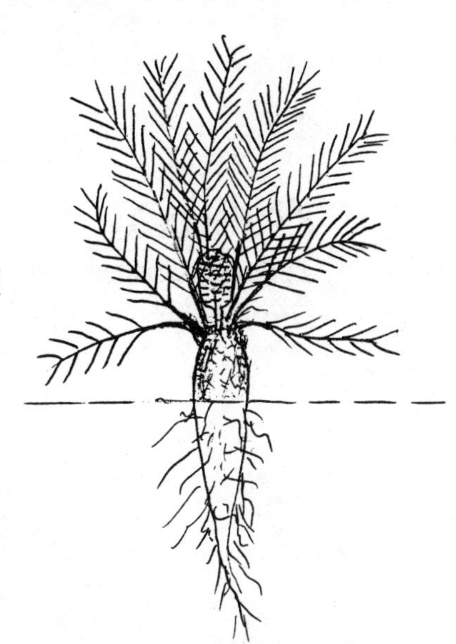

GROWTH HABIT: Coarse woody evergreen, somewhat fern-like, from woody stem that extends well down into soil

LEAVES: All arising from base, each consisting of 6-18 pairs of linear lateral divisions

FLOWERS: None, a cone producer, male and female on different plants

FRUITS: Woody or leathery cone 2-4″ long

SEEDS: Several and rather large

LIGHT: Lots of sun

SOIL: Sandy

MOISTURE: Dry to moist

SEASONAL ASPECT: None

ZONE: 1, and only the lower part

USES: Specimen, for unusual or tropical effect

CARE: None

PROPAGATION BY: Seeds

NATIVE OF: s.e. United States

OTHER: The large mostly underground stem is starchy and was extensively used by Indians for food. Florida Arrowroot, sometimes seen on food store shelves, is from this source. *Z. floridana,* similar but with smaller stem.

Zelkova

Zelkova serrata

Elm Family

GROWTH HABIT: Broadly spreading deciduous tree, somewhat Elm-like
LEAVES: Alternate, short petioled, toothed, somewhat rough above, becoming progressively smaller, bark from twig tips, 1-4″ long
FLOWERS: Greenish and inconspicuous
FRUITS: 1/4″ long, not winged, curiously shaped
SEEDS: 1 per fruit
LIGHT: Full sun
SOIL: Tolerant of most soils
MOISTURE: Moist to dry
SEASONAL ASPECT: None
ZONE: 1, 2
USES: Specimen tree, street side
CARE: None
PROPAGATION BY: Cuttings, seeds
NATIVE OF: Japan, E. Asia to the Caucasus
OTHER: Merits greater use where small trees are desired.

INDEX

SCIENTIFIC NAME INDEX

COMMON NAME INDEX

Common Name Index

Common Name Index